MASS TRANSFER

J. A. WESSELINGH
Professor of Chemical Engineering
University of Groningen, The Netherlands

R. KRISHNA
Professor of Chemical Engineering
University of Amsterdam, The Netherlands

ELLIS HORWOOD
NEW YORK   LONDON   TORONTO   SYDNEY   TOKYO   SINGAPORE

First published in 1990 by
ELLIS HORWOOD LIMITED
Market Cross House, Cooper Street,
Chichester, West Sussex, PO19 1EB, England

A division of
Simon & Schuster International Group
A Paramount Communications Company

Typeset by Ellis Horwood Limited
Printed and bound in Great Britain
by Hartnolls, Bodmin, Cornwall

British Library Cataloguing in Publication Data

Mass Transfer
J. A. Wesselingh and R. Krishna
(Ellis Horwood series in Chemical Engineering)
CIP catalogue record for this book is available from the British Library
ISBN 0–13–553165–9 (Library Edition)
ISBN 0–13–553025–3 (Student Paperback Edition)

Library of Congress Cataloging-in-Publication Data available

MASS TRANSFER

Ellis Horwood Series in
CHEMICAL ENGINEERING
Series Editor: Dr DAVID SHARP, OBE, former General Secretary,
Institution of Chemical Engineers, London

THERMODYNAMICS FOR CHEMISTS AND CHEMICAL ENGINEERS
M.H. CARDEW
ENVIRONMENTAL BIOTECHNOLOGY
C. F. FORSTER and D.A.J. WASE
POWDER TECHNOLOGY AND MULTIPHASE SYSTEMS:
Gas Permeametry and Surface Area Measurement
G. J. I. IGWE
MAJOR CHEMICAL HAZARDS
V. MARSHALL
BIOLOGY AND BIOCHEMISTRY FOR CHEMISTS AND CHEMICAL ENGINEERS
W. J. MITCHELL and J. C. SLAUGHTER
INSTRUMENTATION AND AUTOMATION IN PROCESS CONTROL
M. J. PITT and P. E. PREECE
MASS TRANSFER
J. A. WESSELINGH and R. KRISHNA
BIOLOGICAL TREATMENT OF WASTE-WATER
M. WINKLER

Ellis Horwood Limited also publishes related titles in
APPLIED SCIENCE AND INDUSTRIAL TECHNOLOGY
INDUSTRIAL CHEMISTRY
BIOLOGICAL SCIENCES

Table of contents

1

Beginning ...

WHO SHOULD READ THIS BOOK?

This book is about the mass transfer processes which are really important, but which are neglected in most textbooks:

- those with three or more species and
- those with more than one driving force, such as centrifugal, electrical or pressure gradients.

If you want to know more about these subjects, but find existing texts too difficult, then this is the book for you. Also, if you already understand the intricacies of multicomponent mass transfer, you may find it enjoyable to see how far you can get with simple means.

The book assumes that you have a working knowledge of:

- thermodynamics and phase equilibria; chemical potentials, enthalpies, activity coefficients, partial molar volumes and distribution coefficients,
- transport phenomena: simple mass balances, binary diffusion and mass transfer coefficients.

If you are not too sure, do not despair. We will introduce these concepts in a fairly leisurely manner. Because there are many new ideas to get used to, we have tried to avoid mathematical complexity. For the greater part of the text you do not need more than the ability to solve three linear equations with three unknowns. Indeed, we hope to show that a large part of multicomponent mass transfer can be applied with pencil, paper and a calculator. Of course you will need a computer for larger problems, but not to obtain a first understanding.

WHAT THIS BOOK COVERS

Textbook treatments of mass transfer traditionally start with binary mixtures ... and stop there. However, there are few real processes involving only binary mass transfer. Even a distillation usually has more components. In membrane processes the simplest cases involve three species: the two to be separated and the membrane itself. Unfortunately, multicomponent mixtures possess properties which are not found even qualitatively in binary mixtures. Also the binary approach cannot be extended easily to mixtures with more than two components. The binary start is a dead-end alley.

This book takes the approach that species move with respect to each other owing to their potential gradients. The rate of movement is restricted by friction between the species. This method was already brought forward by Maxwell and Stefan more than a century ago. It has not caught on, probably because the mathematics are thought to be difficult. This is not really a problem however:

- there are simple approximations to the solutions of the equations. This is the approach taken in this book.
- the computer and numerical techniques now make 'exact' calculations much easier.

Taking potential gradients allows the incorporation of different driving forces:

- composition (or more precisely activity) gradients,
- electrical potential gradients,
- pressure gradients,
- centrifugal fields

and so on. Also thermodynamics and transport phenomena now become one subject. Equilibrium is simply the situation where the driving forces have disappeared.

The friction approach to interactions between the species allows any number of components to be handled in a consistent manner. More components only means more and longer equations.

With these starting points almost any mass transfer process can be described. Examples in this book cover:

- multicomponent distillation, absorption and extraction,
- multicomponent evaporation and condensation,
- heterogeneous catalysis
- sedimentation and ultracentrifugation,
- filtration,
- electrolysis,
- dialysis,
- pervaporation,
- electrodialysis,
- membrane gas separations,
- reverse osmosis,
- ultrafiltration,
- ion exchange and adsorption.

The examples treat diffusion in gases, in liquids, in electrolyte solutions, in porous media and in swollen polymers. The book includes some data for estimating multicomponent diffusivities and mass transfer coefficients.

A major limitation of the book is that it only covers situations with a transfer resistance in one of the phases. Such a resistance will be a building block for the simulation of separation or reactor equipment. The reader must be prepared to incorporate the equations into his own simulations. The approximations used should be sufficiently accurate for most engineering applications.

GUIDELINES TO THE READER

This text was written to accompany overhead transparencies in a course on multicomponent mass transfer. So the figures are quite important. Indeed you can get a fair impression of the contents by glancing through the figures and reading the summaries after every chapter (except Chapters 14 and 15). Not all chapters are equally important. If you are convinced that you know everything about mass transfer (as we used to be!) you should read Chapter 2. It may contain a few surprises. Chapters 3 and 4 are the basis for the rest and are essential. Binary examples to start with are given in Chapter 5. Your first multicomponent calculations are at the beginning of Chapter 6. At that point you should know enough to start making your own choices. Of course not every reading sequence is equally handy. For example, if you want to study electrodialysis you should take Chapters 11 (electrolyte solutions) and 12 (membrane processes) first. Otherwise it is up to you.

We have given exercises at several points in the text. Most of these are small and only require pencil, paper and a calculator. We advise you to try them. A set of worked examples is also included. Many of these are much larger; they may require several hours work. We have solved them using 'MathCAD', but other packages should do a similar job.

2

Is something wrong?

THE STARTING POINT

We expect your working knowledge of mass transfer to be something
like that summarized in Fig. 2.1. Once these were also the only tools

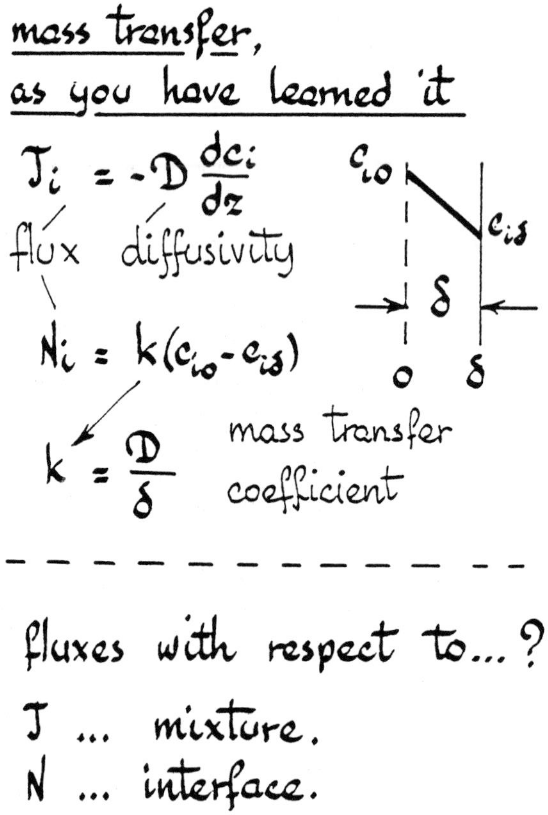

<u>mass transfer,</u>
<u>as you have learned it</u>

$$J_i = -D \frac{dc_i}{dz}$$

flux diffusivity

$$N_i = k(c_{i0} - c_{is})$$

$$k = \frac{D}{\delta} \quad \text{mass transfer coefficient}$$

- - - - - - - - - - - - - - - -

fluxes with respect to...?

J ... mixture.

N ... interface.

Fig. 2.1

we had. So let us have a look at them. There are two 'laws':

- the flux of a species is proportional to its concentration gradient
  and

- the flux is proportional to a concentration difference times a mass transfer coefficient.

You may regard the second 'law' as an integration of the first over a thin film. It is a rough model for a mass transfer resistance.

The two laws are handy little formulae for describing simple mass transfer problems. In passing we note that the fluxes and concentrations used here are in molar units ($mol\,m^{-2}\,s^{-1}$ and $mol\,m^{-3}$). This implies that the diffusivity has units of $m^2\,s^{-1}$ and the mass transfer coefficient of $m\,s^{-1}$ (a velocity). We distinguish two kinds of fluxes. One (J) with respect to the (moving) mixture. The other (N) with respect to an interface. 'J' is the more fundamental of the two, but 'N' is the one usually required by the engineer. More about this later.

Conventional mass transfer in a binary mixture of gases (Fig. 2.2)

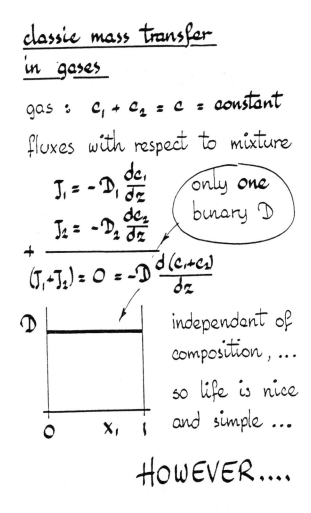

classic mass transfer in gases

gas : $c_1 + c_2 = c = $ constant

fluxes with respect to mixture

$$J_1 = -\mathcal{D}_1 \frac{dc_1}{dz}$$

$$J_2 = -\mathcal{D}_2 \frac{dc_2}{dz}$$

(only one binary $\mathcal{D}$)

$$+ \overline{\qquad\qquad}$$

$$(J_1 + J_2) = 0 = -\mathcal{D}\frac{d(c_1 + c_2)}{dz}$$

independent of composition, ...

so life is nice and simple ...

HOWEVER....

Fig. 2.2

is especially simple. If pressure and temperature in the gas are assumed constant, the total molar concentration is also constant. By definition the sum of the fluxes with respect to the mixture is zero. If you then write down the flux equations for the two components it is immediately clear that there is only one binary diffusion coefficient. Also experiment (and the kinetic theory of gases) tells us that this diffusivity is a constant (that is: independent of composition, not of pressure and temperature). The simplicity of the above rules might lead you to believe that this is all there is to mass transfer. However, that is not so as we shall see in the following examples.

THREE GASES

Look at the experiment in Fig. 2.3. There are two equal glass bulbs

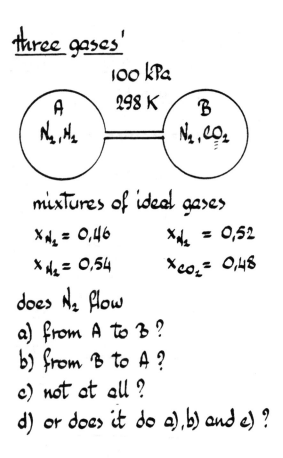

three gases'

100 kPa

298 K

A N_2, H_2 B N_2, CO_2

mixtures of ideal gases

$x_{N_2} = 0.46$ $x_{N_2} = 0.52$

$x_{H_2} = 0.54$ $x_{CO_2} = 0.48$

does N_2 flow

a) from A to B ?

b) from B to A ?

c) not at all ?

d) or does it do a), b) and c) ?

Fig. 2.3

filled with mixtures of ideal gases. The left bulb consists of hydrogen and nitrogen, and the right bulb of carbon dioxide and nitrogen. The amounts of nitrogen in the two bulbs only differ slightly. Both bulbs are at the same pressure and temperature. At a certain moment they are connected by a capillary. The capillary is fairly narrow, say with a diameter of one millimetre, but otherwise nothing special. Gases start diffusing from one bulb to the other. Before you read on we would like to ask you to think a moment about the questions given in the figure, and to make your own decision on which answer you choose.

— Have you answered the questions? —

The results of the experiment are shown in Fig. 2.4. The behaviour of

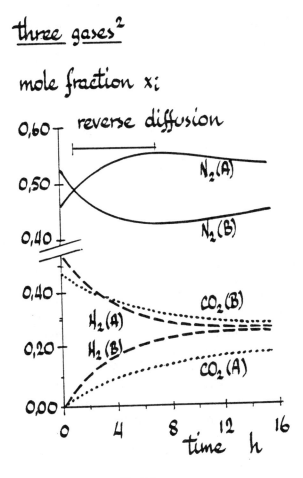

Fig. 2.4

hydrogen and carbon dioxide (bottom part of the figure) is as expected. Their compositions change monotonically in such a way that after a few days the amounts in the two bulbs will have become equal. Hydrogen moves more rapidly than carbon dioxide. Nitrogen (note the difference in the composition scale) behaves quite differently. Initially it diffuses from the high concentration (bulb B) to the low concentration, and the two concentrations become equal after about one hour. However, nitrogen keeps on diffusing in the same direction, now against its concentration gradient. The gradient keeps increasing up to about eight hours after the start of the experiment. Only after that do the two bulbs gradually go back to equal compositions.

With the conventional mass transfer theory of the first two figures in mind you will probably find it very difficult to understand what is going on. So just try to forget these for a moment and try a different viewpoint. It is fairly obvious why hydrogen is going from left to right. There is far more hydrogen in the left bulb than in the right one, and the random thermal motions of the molecules will on average cause them to move to the right. The same mechanism causes carbon dioxide to move to the left. Now it looks as if nitrogen is being dragged along by the carbon dioxide. This is understandable: you would expect more friction between the heavy carbon dioxide molecules and nitrogen than between nitrogen and hydrogen. At least initially the movement of nitrogen is mainly determined by the carbon dioxide and hydrogen gradients, and not by its own gradient (which is rather small). There is no such mechanism in the conventional laws of mass transfer.

TWO CATIONS

This experiment involves a membrane which is permeable to cations, but not to anions and water (Fig. 2.5). We bring a dilute solution of sodium chloride in the right compartment and a much more concentrated solution of hydrochloric acid in the left compartment. You would expect the sodium ions to diffuse from the right to the left until the two concentrations have become equal. They do indeed go in that direction. However, they may go on until their concentration in the left compartment is many times higher than in the right one!

The experiment has similarities with the previous one. The explanation, however, is rather different. Hydrogen ions diffuse through the membrane to the right. (Remember that the chloride ions cannot.) This causes a small positive excess charge in the right compartment. The resulting electrical gradient forces the sodium ions to the left and restricts the amount of hydrogen that can be transferred. Again there is no such mechanism in conventional mass transfer.

TWO GASES AND A POROUS PLUG

As a last example (Fig. 2.6) we consider a porous plug. It is a plug with fine openings (you might think of compressed cotton wool), but otherwise inert. On one side there is helium, on the other argon, both

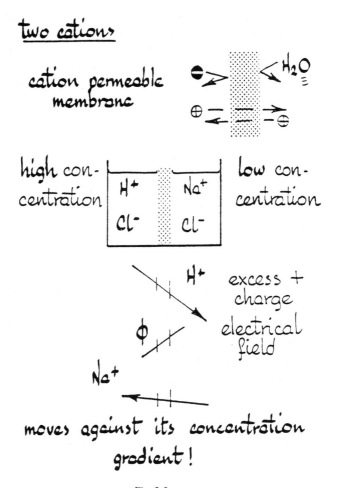

Fig. 2.5

at exactly the same pressure and temperature. Such a situation can be maintained by having flows of pure helium and pure argon past the ends of the plug, passing out into the open. The concentration gradients are equal. A careless application of Fig. 2.2 might let you expect that the fluxes should also be equal.

In reality, experiment shows that the helium flux is about three times higher than the argon flux! If you want to understand this you should realize that Fig. 2.2 only tells you something about the movement of helium and argon with respect to each other. It tells you nothing about the movement of the mixture with respect to the plug. This movement is balanced by friction between the two gases and the plug. To let the two friction terms cancel, helium must have the higher velocity. There is no interaction with any plug in the conventional analysis of mass transfer; clearly you should take the plug as a component of the mass transfer system.

By applying a pressure difference (Fig. 2.7) you can equalize the

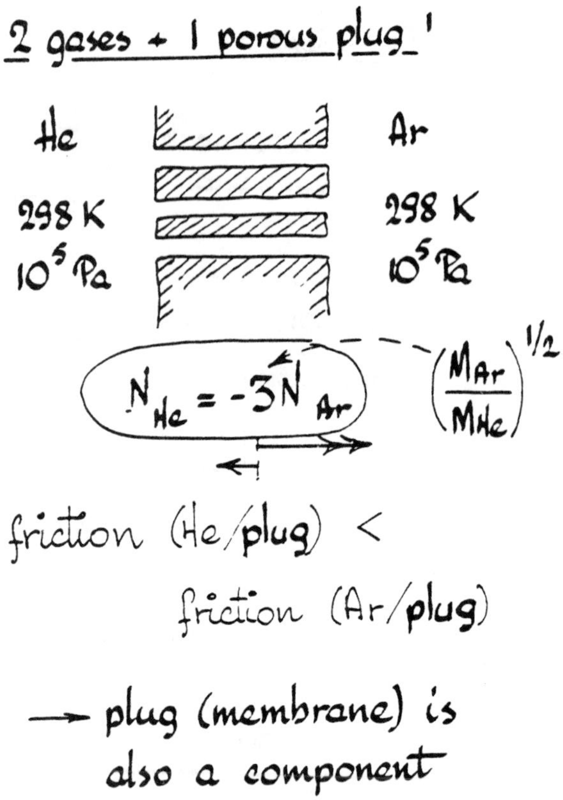

2 gases + 1 porous plug !

He   [diagram]   Ar

298 K   [diagram]   298 K

10^5 Pa   [diagram] 10^5 Pa

$$N_{He} = -3 N_{Ar} \qquad \left(\frac{M_{Ar}}{M_{He}}\right)^{1/2}$$

friction (He/plug) <

friction (Ar/plug)

→ plug (membrane) is
also a component

Fig. 2.6

two fluxes. The viscous flow due to the pressure gradient increases the argon flux and decreases the helium flux. This is analogous to the effect of the electrical gradient in the previous example. The pressure difference required depends on the structure of the plug: the finer the pores, the larger the pressure difference.

In the three-gases problem we saw earlier, the capillary is much wider. Even so, rapid diffusion of hydrogen does cause a minute pressure difference. The transport of nitrogen against its gradient is in part due to the resulting viscous flow.

SUMMARY

In this second chapter we have briefly reviewed conventional mass transfer. We have then applied it to a few examples and seen that it can give rise to large problems.† The examples suggest that a better approach to mass transfer would have to take the following pheno- mena into account:

† We hope that you realize that there is nothing special in these examples. They deal with mixtures of ideal gases, inert plugs and simple electrolyte solutions, all under quite normal conditions.

2 gases + 1 porous plug[2]

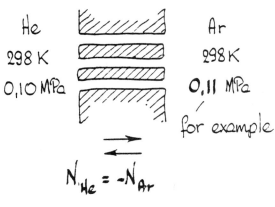

He   Ar
298 K   298 K
0,10 MPa   0,11 MPa
                            / for example

$$N'_{He} = -N'_{Ar}$$

main reason : **viscous flow**

$$—\Delta p—\rightarrow$$

retards He ,
accelerates Ar

Fig. 2.7

- friction between each pair of components, including any solid matrix,
- the occurrence of other driving forces than only composition gradients. You can think of electrical and pressure gradients.
- in heterogeneous media (solid matrices) you should take viscous flow into account.

We will be working these ideas out in the next two chapters.

3

Driving forces

TOWARDS A NEW THEORY

The elements of the new theory are summarized in Figs 3.1–3.3. The resistance for mass transfer is thought to be located in a 'film' in which there is no eddy motion. This is a good description of a membrane. It is only a rough model for transfer through fluid/fluid or fluid/solid interfaces. However, it is adequate for many engineering applications. The film is usually very thin; a few orders of magnitude are indicated in Fig. 3.1.

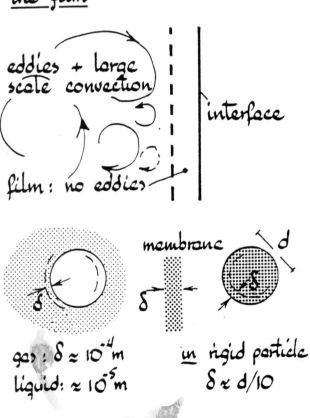

Fig. 3.1

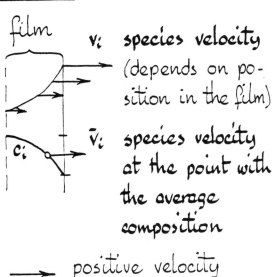

velocities

film

v_i **species velocity** (depends on position in the film)

$\bar{v}_i$ **species velocity at the point with the average composition**

c_i

→ positive velocity

Fig. 3.2

We will not (immediately) focus on the fluxes of the species being transferred, but on their velocities (Fig. 3.2). The velocity we are talking about is the average velocity of the species towards the interface. This value can differ greatly: from below one micrometre per second in solid matrices to around ten centimetres per second in gases. Velocities in liquids lie between these values: usually around a tenth of a millimetre per second. The species velocity should not be confused with the thermal velocity of individual molecules. This has a value of several hundred metres per second. However, the thermal movements are for the greater part random. This random part gives no contribution to the species velocity.

In stationary transport through a flat film a species will have the same flux at each depth. Its velocity must then vary with depth. It will be smallest where the concentration is highest and vice versa. In the new theory we will not pay attention to such variations. We will focus on obtaining an estimate of the velocity at the average composition of the species in the film. For brevity this will be called _the_ velocity. Note that a velocity is taken as positive when a species moves from left to right.

The species velocity is determined by several factors:

* the movement of the mixture as a whole,
* the potential gradient of the species and
* friction of the species with its surroundings.

The first point should be clear; a species tends to move along with the mixture.

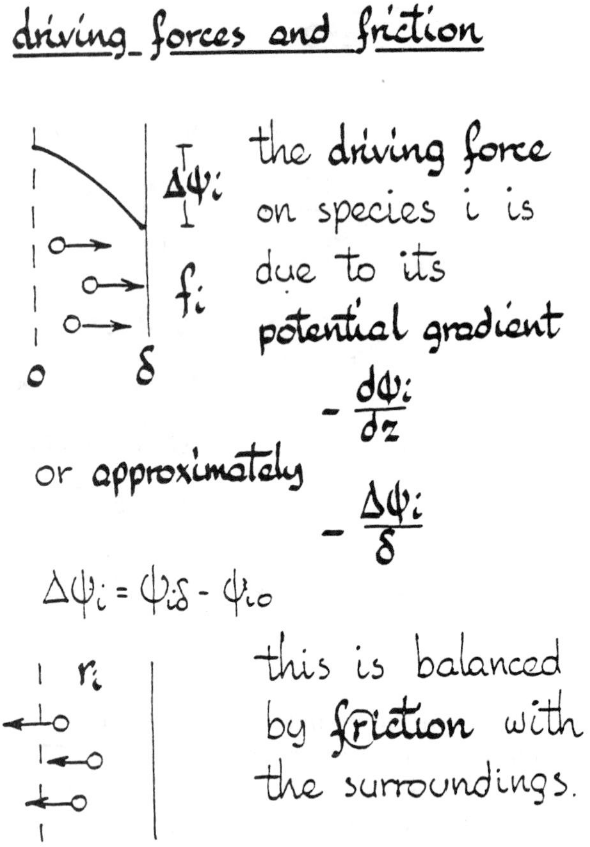

driving forces and friction

$$\Delta \psi_i \qquad \text{the driving force}$$
$$f_i \qquad \text{on species } i \text{ is}$$
$$\text{due to its}$$
$$\text{potential gradient}$$
$$-\frac{d\psi_i}{dz}$$

or approximately

$$-\frac{\Delta \psi_i}{\delta}$$

$$\Delta \psi_i = \psi_{i\delta} - \psi_{io}$$

this is balanced
by friction with
the surroundings.

Fig. 3.3

The potential gradient of a species will tend to move the species with respect to the mixture (Fig. 3.3). We will usually approximate this gradient by the potential difference over the film, divided by the film thickness. The potential difference is taken as that at the right side of the film minus that at the left side. With this convention, a negative potential difference causes a positive driving force. We will come back to the driving force a little further on.

The rate at which the species moves through the mixture is limited by friction with the other species. This friction is the subject of Chapter 4. In the motion of a molecular species its driving force and total friction can be taken as equal but opposite. Acceleration forces on the species are usually of no importance. (Exceptions are in sonic flows and shock waves from detonations.)

A SIMPLE DRIVING FORCE: GRAVITY

Just to repeat a few elementary concepts, let us consider gravity (Fig. 3.4). This is an example of an 'external force'. Suppose you raise

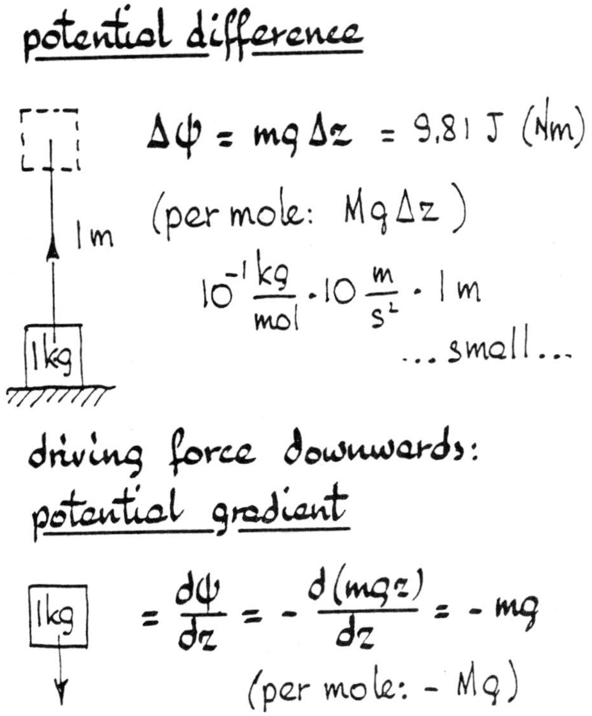

potential difference

$$\Delta\varphi = mg\,\Delta z = 9.81 \text{ J (Nm)}$$

$$(\text{per mole: } Mg\,\Delta z)$$

$$10^{-1}\frac{kg}{mol}\cdot 10\,\frac{m}{s^2}\cdot 1\,m$$

$$\ldots\,small\ldots$$

driving force downwards: potential gradient

$$= \frac{d\varphi}{dz} = -\frac{d(mgz)}{dz} = -mg$$

$$(\text{per mole: } -Mg)$$

Fig. 3.4

a mass slowly ('reversibly') in the earth's gravity field. The work you perform is equal to the potential increase of the mass. The gravitational force on the mass is the negative of the potential gradient. It is seen to have a value of the order of one Newton per mole. This is a very small force when considering molecular motion.

There are other, more important external force fields, such as those in centrifuges, the support forces of solid matrices in membrane processes and electrical fields. We will come back to these in Chapter 10 (external forces), Chapter 11 (electrolyte solutions) and Chapters 12–16 (membrane processes).

THE COMPOSITION DRIVING FORCE

Every substance not only has a gravitational potential, but also a chemical potential. This potential is usually lower if the substance is mixed with other components. The work required to (reversibly) separate one mole of i from a large amount of a mixture is the contribution of the composition to the chemical potential (Fig. 3.5). It typically has a value of a few thousand Joules per mole.

As you can see, the chemical potential is a logarithmic function of the 'activity' of the species. This in turn is the product of its 'activity coefficient' and its mole fraction. Gases and liquid mixtures of similar

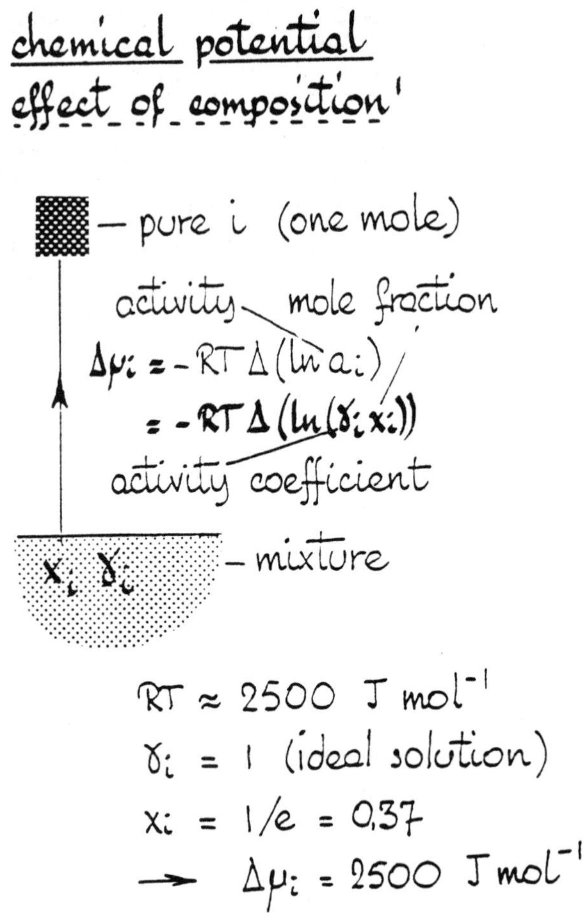

<u>chemical potential</u>
<u>effect of composition</u>'

■ — pure i (one mole)

   activity   mole fraction

$\Delta \mu_i = - RT \Delta (\ln a_i)$

   $= - RT \Delta (\ln(\gamma_i x_i))$

activity coefficient

$x_i \; \gamma_i$ — mixture

$$RT \approx 2500 \ J \ mol^{-1}$$
$$\gamma_i = 1 \ (ideal \ solution)$$
$$x_i = 1/e = 0.37$$
$$\rightarrow \quad \Delta \mu_i = 2500 \ J \ mol^{-1}$$

Fig. 3.5

components form 'ideal' solutions. The activity coefficient then has a value of one and can be omitted. This simplifies formulae, and we will often assume ideality in our examples. Non-ideality is considered in detail in Chapter 8.

For ideal solutions, a handy little approximation of the composition effect is given in Fig. 3.6. The approximation holds over quite wide spans of compositions. The complete description of mass transfer that we develop in the next chapter also contains approximations. It has been found that the approximate formula for the effect of composition change gives better results in the final equations than the exact formula of the previous figure. Apparently there are compensating errors.

To become accustomed to the previous formulae, you might well try the exercise in Fig. 3.7. It concerns the force driving carbon dioxide out of beer (or cola, if you prefer) into bubbles. The problem should only take you a minute or so to solve.

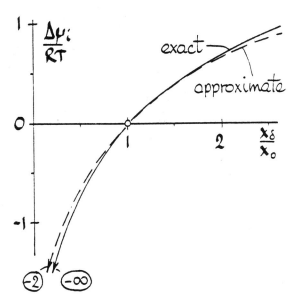

$$\text{\underline{effect of composition}}^2$$
$$\underline{\text{on approximation}}$$

$$\Delta\mu = RT \ln \frac{x_\delta}{x_0} \approx RT \frac{(x_\delta - x_0)}{(x_\delta + x_0)/2}$$

$$\boxed{\Delta\mu_i \approx RT \frac{\Delta x_i}{\bar{x_i}}} \quad \text{ideal solution}$$

Fig. 3.6

Try it.

You should have obtained a value of about five hundred thousand ton force per kilogram of carbon dioxide. 'Every kilogram feels the weight of a supertanker!' This illustrates the enormous size of molecular forces in mass transfer.

EFFECT OF A TEMPERATURE GRADIENT

The formula in Fig. 3.6 remains correct when there is a temperature gradient. For the temperature, you should take the average in the film. Do *not* put the temperature after the difference sign! This is difficult to explain here: you will have to take it for granted. It does not mean that a temperature gradient has no influence on mass transfer. We come back to this in Chapter 7.

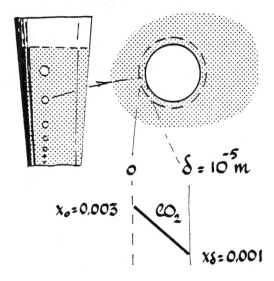

forces in a glass of beer

$\delta = 10^{-5} m$

$x_0 = 0,003$ CO_2

$x_\delta = 0,001$

what is the force per kg of CO_2?

$M_{CO_2} = 44$ g mol^{-1}

Fig. 3.7

SUMMARY

In this chapter we have seen the following.

- The resistance to mass transfer between two phases is thought to be located in thin 'films' next to the phase interface.
- Species are driven through the film by their potential gradients.
- A simple potential gradient is due to gravity. It turns out to be rather unimportant (except for very large molecules).
- Much more important is the gradient of the chemical potential. This is caused by gradients in the mole fractions of the species. In non-ideal solutions also gradients in the activity coefficients of the species are important.
- On a molecular scale, forces due to potential gradients can be extremely large.
- We have noted that the temperature does not enter into the 'difference' of the chemical potential.

You should memorize the approximate form for the driving force in Fig. 3.6.

4

Friction

In this chapter we will first have a look at friction between the species. Driving forces and friction will then be combined to yield our transport relation. We will also see how this relation can be 'tied' to the surroundings.

SPHERES IN A LIQUID

Driving forces can be very large. Even so, species usually move slowly; they 'feel' a very large resistance. Why is this so?

In Fig. 4.1 a simple model is shown of the behaviour of a dilute solution of tiny spheres in a liquid. The driving force is again their chemical potential gradient. The resistance is caused by the hydro-dynamic drag on the spheres as they move through the liquid. The drag on a single sphere is not large, but the total friction is enormous. This is because of the extremely large number of spheres in one mole. Note that the model predicts the friction to be proportional to the difference between the species velocities.

You can rearrange the transport equation to obtain a definition of the 'diffusivity' (Fig. 4.2). This is a Maxwell–Stefan diffusivity. In this simple example it is identical to the Fick diffusivity that you already know. The diffusivity is inversely proportional to the diameter of the spheres and the viscosity of the liquid.

For transport through a film we make a difference approximation of the previous equation (Fig. 4.3). This is rearranged in a dimension-less form, in which you can, however, still recognize:

$$-(\text{driving force on '2'}) = (\text{friction on '2'})$$

The transport relation contains a constant with the dimension of a velocity. This constant is a mass transfer coefficient. It is a close relative of the mass transfer coefficients we see in conventional mass transfer theory.

a simple model of friction
in liquids

"1" : the liquid

"2" : a spherical solute

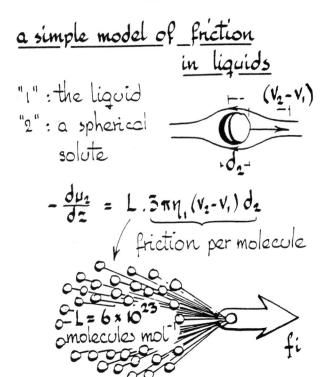

$$-\frac{d\mu_2}{dz} = L \cdot \underbrace{3\pi\eta_1 (v_2 - v_1) d_2}_{\text{friction per molecule}}$$

$L = 6 \times 10^{23}$ molecules mol^{-1}

* friction is large because there are so many molecules per mole...
* friction is proportional to $(v_2 - v_1)$

Fig. 4.1

"sphere 'in liquid" model[2]

$$-\frac{d}{dz}\left(\frac{\mu_2}{RT}\right) = \frac{v_2 - v_1}{Ð_{12}}$$

Maxwell-Stefan diffusivity

$$\boxed{Ð_{12} = \frac{RT}{L \, 3\pi\eta_1 d_2}} \quad m^2 s^{-1}$$

Fig. 4.2

<u>transport relation</u> '

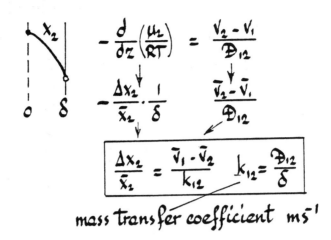

$$-\frac{d}{dz}\left(\frac{\mu_2}{RT}\right) = \frac{\bar{v}_2 - \bar{v}_1}{\mathcal{D}_{12}}$$

$$-\frac{\Delta x_2}{\bar{x}_2} \cdot \frac{1}{\delta} \qquad \frac{\bar{v}_2 - \bar{v}_1}{\mathcal{D}_{12}}$$

$$\boxed{\frac{\Delta x_2}{\bar{x}_2} = \frac{\bar{v}_1 - \bar{v}_2}{k_{12}} \qquad k_{12} = \frac{\mathcal{D}_{12}}{\delta}}$$

mass transfer coefficient $m s^{-1}$

Fig. 4.3

<u>friction</u>

$$\left(\begin{array}{c}\text{friction between } j \text{ and } i \\ \text{per mole of } i\end{array}\right) \sim$$

$$\left(\begin{array}{c}\text{concentration} \\ \text{of } j\end{array}\right)\left(\begin{array}{c}\text{velocity difference} \\ \text{between } j \text{ and } i\end{array}\right)$$

$$\boxed{= \bar{x}_j \frac{\bar{v}_j - \bar{v}_i}{k_{ij}}}$$

for a matrix or membrane (m)

$$\frac{\bar{x}_m}{k_{im}} \leftarrow \frac{1}{\mathbb{k}_i}$$

membrane coefficient

Fig. 4.4

As an exercise we suggest that you estimate the diffusivity of spheres of one nanometre in a liquid such as water. This has a viscosity of 1 mPa.s. Take the film thickness from Fig. 3.1 to estimate a mass transfer coefficient.

THE TRANSPORT EQUATION

The idea behind the previous model is easily extended to more general cases with any number of components (Fig. 4.4). Here we look at a certain species i and its friction with another species j. This is assumed to be proportional to the velocity difference between the species. It seems that this is a good approximation; only a few cases are known where it does not apply.[†]

We also take the friction with j to be proportional to the local mole fraction of j. This makes the transport coefficient less dependent on composition. (We will often regard it as a constant.)

In membranes it is sometimes difficult to assign a mole fraction to the solid matrix. Then it can be handier to define a membrane coefficient as shown in the figure.

The driving force on i will be balanced by the friction with all other species. This yields our transport equation (Fig. 4.5). There is one

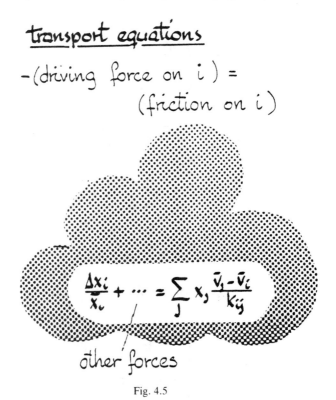

transport equations

$$-(\text{driving force on } i) =$$
$$(\text{friction on } i)$$

$$\frac{\Delta x_i}{x_i} + \cdots = \sum_j x_j \frac{\bar{v}_j - \bar{v}_i}{k_{ij}}$$

other forces

Fig. 4.5

† An example is in the non-steady penetration of small molecules into a glassy polymer. Here the transport is governed by the rate at which the polymer uncoils. This is not well described by a friction model.

such relation for each component (however, do read on to the next paragraph). The equation is a difference form of what is known as the Maxwell–Stefan equation. It plays a central role in the rest of this book, and we advise you to memorize it.

The corresponding differential equation is shown in Fig. 4.6. This equation is rather difficult to solve. In this introductory text we will therefore stay with the difference approximation. In our experience this contains all the main features of the differential equation. Also its accuracy is adequate for many engineering applications.

$$\text{the differential equation:}$$

$$\frac{1}{x_i} \cdot \frac{dx_i}{dz} = \sum_j x_j \frac{v_j - v_i}{\mathcal{D}_{ij}}$$

$$\text{(not used in this book).}$$

Fig. 4.6

If you have (a lot of) time you may wish to attack problems in a more fundamental way using the complete differential equation. We will give you references on this at the end of this book.

MASS TRANSFER COEFFICIENTS

With multicomponent systems there will be friction between i and several other components. The total friction on i is the sum of these values. To estimate them we need several mass transfer coefficients: one for each pair of interactions.

Fig. 4.7 gives a rough idea of the ranges of mass transfer coefficients encountered. For gases they are usually between one centimetre and one metre per second. For liquids, they are a factor of a thousand lower. Transport coefficients in porous media are typically an order of magnitude lower than in free solution. In tight polymer matrices they may be two orders of magnitude lower. Far lower coefficients are encountered in solids used for packaging, but these are not of interest in mass transfer operations.

THE BOOTSTRAP RELATION

If you look again at our transport relation, you will see that it only contains velocity differences. There is no absolute velocity in the equation. It only tells you about the relative velocities in the mixture, nothing about the mixture as a whole. Our transport equations are floating relations as you might have seen from the shape of their

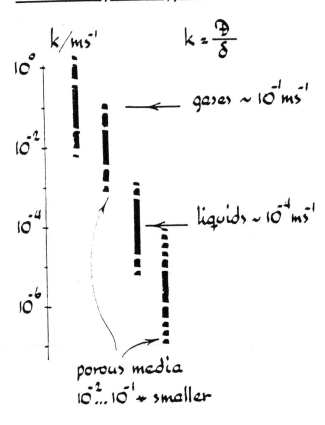

Fig. 4.7

circumference in Fig. 4.5. As you can see in Fig. 4.8 this implies that you must drop one of the transport relations. To obtain the absolute velocities, you must then add a relation of a completely different kind. This relation ties the problem down to real earth and gives the velocities with respect to the frame of reference that you choose. It is therefore known as the bootstrap relation (Fig. 4.9).

In membrane processes the membrane is commonly assumed to be stationary. This yields the bootstrap relation at the bottom of the figure. In other cases the bootstrap relation is not always as obvious. We will, however, give many examples in the coming chapters. The bootstrap relation can usually be written in the form of a linear relation between the velocities.

FLUXES

The process engineer is not usually interested in species velocities but in the fluxes through an interface. The flux is equal to the product of only _relative velocities_ ...

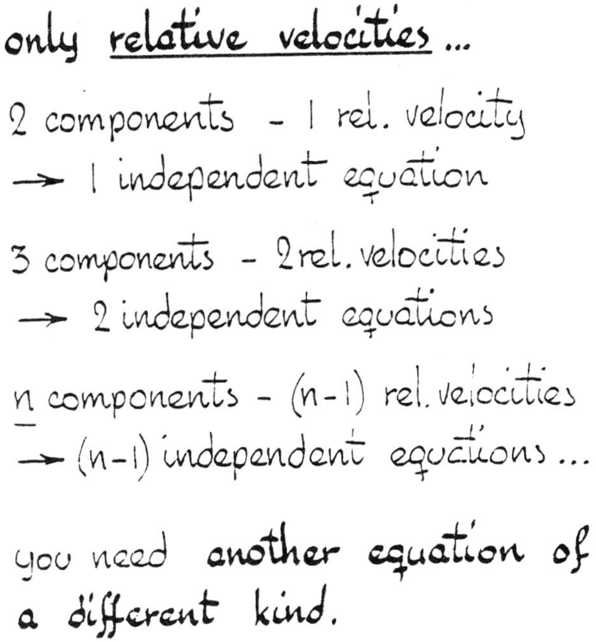

2 components - 1 rel. velocity

→ 1 independent equation

3 components - 2 rel. velocities

→ 2 independent equations

n components - (n-1) rel. velocities

→ (n-1) independent equations ...

you need **another equation of a different kind.**

Fig. 4.8

the species velocity and the species concentration (Fig. 4.10). You may substitute the fluxes in the transport equations if you wish. This is often handy in real problems. However, it does somewhat obscure the meaning of the terms. In most of this book we have chosen to keep velocities in the transfer relations and to calculate the fluxes afterwards.

Any viscous flow is easily incorporated in the flux calculations. The two velocities are additive. The viscous velocity is the component of the pore velocity perpendicular to the film. It is not the superficial velocity.

SUMMARY

- In this chapter we have derived the mass transfer relation which will be the backbone of the rest of this book.
- You should by now have memorized this relation.
- The transport relation of a species contains driving forces on the species and friction terms with all other species.
- The friction terms contain mass transfer coefficients which are directly related to a new type of diffusivity.
- The number of independent transport relations is one less than the number of components.
- To obtain velocities with respect to a phase interface you must therefore have an extra 'bootstrap' relation to tie the problem down.

boot-strap relation

"floating" transport relations...

$$\frac{\Delta x_i}{\bar{x}_i} + \cdots = \sum_j \bar{x}_j \frac{\bar{v}_j - \bar{v}_i}{k_{ij}}$$

... have to be tied to surroundings

$$\sum v_i \bar{v}_i = 0$$

example: membrane processes

$$V_m = 0$$

$$(\gamma_m = 1 \quad ; \quad \gamma_{i \neq m} = 0)$$

Fig. 4.9

- Once you know the velocities, fluxes are calculated as the product of the species velocity and the species concentration. Any viscous flow is included in the flux calculations at this point.

velocities and fluxes

$\bar{v}_i$ from transport + bootstrap...

$$N_i = \bar{v}_i \bar{c}_i = \bar{v}_i c \bar{x}_i$$

fluxes with respect to the phase boundary

in heterogeneous media (membranes) include viscous flow:

$$N_i = (\bar{v}_i + u) c \bar{x}_i$$

Fig. 4.10

5

Binary examples

This chapter covers examples with ideal binary mixtures having only a composition gradient as a driving force. Here the advantages of our new approach to mass transfer are not large. However, the examples allow you to get acquainted with the method on small sets of small equations.

STRIPPING

In the first example (Fig. 5.1) ammonia is being stripped from drops of water, into an atmosphere of nitrogen. It is assumed that water evaporation is negligible, and that the bulk concentration of ammonia is very small. We focus on the gas film.

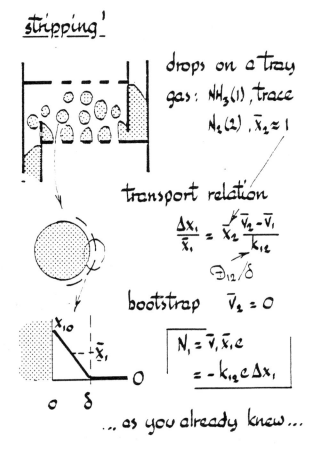

stripping !

drops on a tray

gas: $NH_3(1)$, trace $N_2(2)$, $\bar{x}_2 \approx 1$

transport relation

$$\frac{\Delta x_1}{\bar{x}_1} = \bar{x}_2 \frac{\bar{v}_2 - \bar{v}_1}{k_{12}}$$

$\mathcal{D}_{12}/\delta$

bootstrap $\bar{v}_2 = 0$

$$N_1 = \bar{v}_1 \bar{x}_1 c = -k_{12} c \Delta x_1$$

... as you already knew ...

Fig. 5.1

There are only two components. So there is one independent transport equation. We choose the equation for ammonia. (If you choose that for nitrogen you will see that you get the same equation after a little rearranging.) Nitrogen does not transfer through the interface. So the nitrogen velocity is zero. The nitrogen mole fraction is unity. Inserting all this into the transport relation gives the velocity. From the velocity we compute the flux. This is seen to be equal to the product of the mass transfer coefficient and the concentration difference. In this simple example the new method gives the same result as the classic method, albeit in a roundabout manner.

The only difference in the second example is that the average mole fraction ammonia in the gas film is now one half (Fig. 5.2). Set up the transport relation and bootstrap equation and solve the flux yourself. The answer is given upside down under the figure.

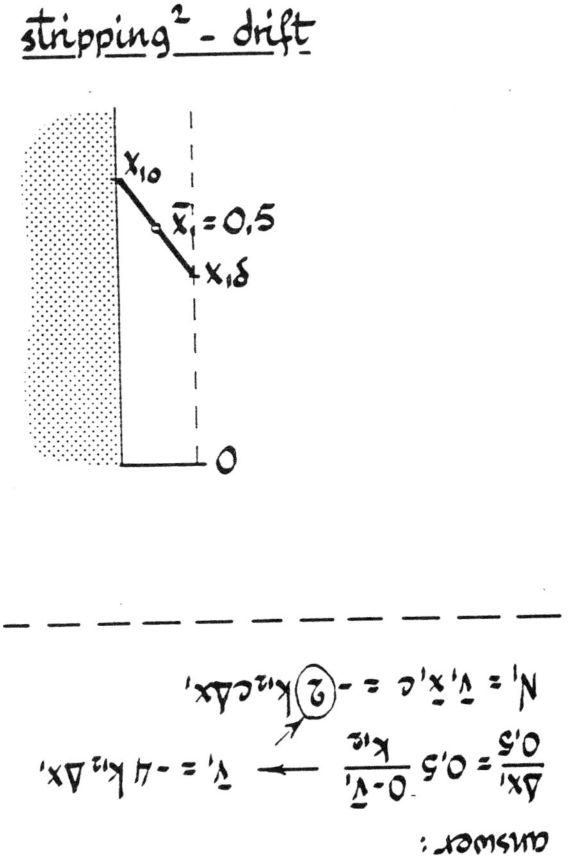
stripping² - drift
x_{10}
$\bar{x}_1 = 0.5$
$x_{1\delta}$
0
answer:
$\frac{0.5}{\Delta x_1} = 0.5\,\frac{0 - v_1}{k_{12}} \rightarrow v_1 = -4\,k_{12}\,\Delta x_1$
$N_1 = v_1 x_1 c = -2\,k_{12}\,c\,\Delta x_1$

Fig. 5.2

As you can see, the flux is now equal to twice the product of mass transfer and concentration difference. This may seem a little surprising, but the same result is also obtained in classical mass transfer.

There, a drift-correction or Stefan-correction is required in this case. The value of the correction factor is indeed two. No such correction is required in the new theory; drift forms an integral part of the equations.

POLARIZATION

A phenomenon which can be important in membrane processes is 'polarization' (Fig. 5.3). Here, water is permeating through a membrane which retains dissolved salt. The water transport causes an increase in salt concentration against the membrane. How large is this increase?

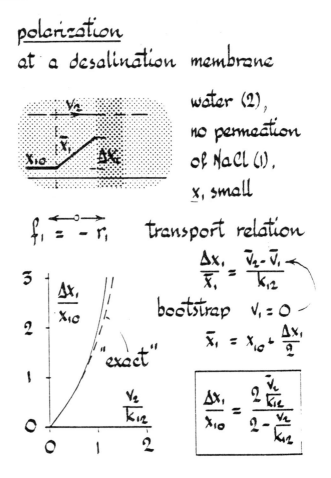

Fig. 5.3

The transport relation for the salt should be obvious. The bootstrap relation is that the salt velocity is zero. This is because the salt cannot pass through the membrane. Working out the equations is trivial. You see that the salt concentration starts rising very sharply as the value of the water velocity approaches the value of the mass transfer coefficient.

An 'exact' solution of the steady-state film model for this case is not difficult to obtain. As you can see, the 'exact' and approximate solutions are very similar.

VAPORIZATION

This (Fig. 5.4) is a case which has puzzled many engineers. A drop consisting of a binary mixture of benzene and toluene is evaporating in a flash vessel. Benzene is the more volatile, and it evaporates that much more rapidly. It has the expected concentration gradient downwards towards the interface. However, this implies that toluene must have a gradient upwards. Yet toluene is evaporating. So it must be diffusing against its concentration gradient. How can this be?

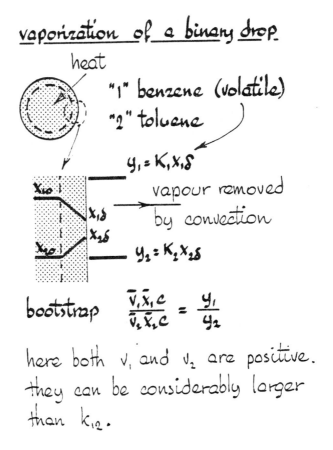

Fig. 5.4

With your knowledge of the Maxwell–Stefan equation, this should be clear, at least in a qualitative manner. The mixture moves towards the interface. Toluene diffuses in the opposite direction, but the diffusive velocity is smaller than the mixture velocity.

The bootstrap relation here is found in the vapour phase. It is assumed that the vapour is removed from the interface by convection (which is a good approximation). The ratio of the two fluxes must then be equal to that of the equilibrium compositions in the vapour at the interface. Modelling this problem with a single steady-state film resistance is rather an oversimplification. However, even this simple model contains most of the essentials of the problem.

If you try this problem along the classic line you will find that you need large drift corrections: a positive one for benzene and a negative one for toluene. There are no such things in the MS description. The equations are just as trivial as in the previous examples; we have not worked them out here.

GASIFICATION OF A CARBON PARTICLE

At sufficiently high temperatures, carbon particles in oxygen react to form carbon monoxide (Fig. 5.5). Oxygen diffuses to the particle

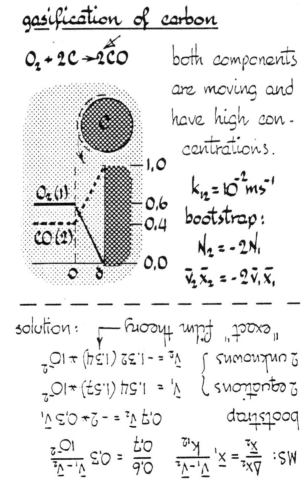

gasification of carbon

$$O_2 + 2C \to 2CO$$

both components are moving and have high concentrations.

$$k_{12} = 10^{-2}\,ms^{-1}$$

bootstrap:

$$N_2 = -2N_1$$

$$\bar{v}_2\bar{x}_2 = -2\bar{v}_1\bar{x}_1$$

solution: — "exact" film theory

2 unknowns { $\bar{v}_e = -1.32\,(1.34) \times 10^{-2}$

2 equations { $\bar{v}_1 = 1.54\,(1.53) \times 10^{-2}$

bootstrap $0.4\,\bar{v}_2 = -2 \cdot 0.3\,\bar{v}_1$

MS: $\dfrac{\Delta x_2}{x_2} = x_1 \dfrac{\bar{v}_1 - \bar{v}_2}{k_{12}}$ $\dfrac{0.6}{0.4} = 0.3\dfrac{\bar{v}_1 - \bar{v}_2}{10^{-2}}$

Fig. 5.5

surface. Its concentration there is very low: it is almost immediately consumed by the reaction. Twice as much carbon monoxide is formed as oxygen is consumed. This diffuses away from the surface into the bulk of the gas. The bootstrap relation is here given by the reaction stoichiometry: the carbon monoxide flux is equal to twice the oxygen flux and opposite in direction.

All data required to calculate the fluxes are given in the figure. You might try it as a last binary exercise. As before, the solution is given upside down in the figure. Note how good the agreement is between the approximate difference method and the exact solution of the differential equation.

SUMMARY

In this chapter we have seen the following.

- The theory used here gives the same results as classic binary mass transfer. (This holds for ideal solutions.)
- In our theory, no drift corrections are required.
- We have seen a number of different bootstrap relations. They are summarized in Fig. 5.6. The only one you have not yet encountered is that for distillation. For a process engineer the assumption of equimolar exchange should not come as a surprise.

some bootstrap relations

① membrane stagnant $v_m = 0$

② bulk stagnant $\bar{v}_2 = 0$
   (absorption)

③ trace stagnant $\bar{v}_1 = 0$
   (membrane polarization)

④ equimolar exchange
   (distillation) $\bar{x}_1\bar{v}_1 + \bar{x}_2\bar{v}_2 = 0$

⑤ interface determined
   (vaporization) $\dfrac{\bar{x}_1\bar{v}_1}{\bar{x}_2\bar{v}_2} = \dfrac{y_1}{y_2}$

⑥ reaction stoichiometry
   $y_1\bar{x}_1\bar{v}_1 + y_2\bar{x}_2\bar{v}_2 = 0$

Fig. 5.6

6

Ternary examples

This is an important chapter. When you have finished it you should have done your first multicomponent calculations.

FROM BINARY TO TERNARY

The extension of the transport relations from binary is straight-forward (Fig. 6.1). There are now two independent transport relations for the three components. In each relation there are two friction terms: for component 1 that between 1 and 2 and that between 1 and 3, etc. Again you need a bootstrap relation to complete the equations. Although there are four friction terms in the two equations, there are only three mass transfer coefficients. The friction between 1 and 2 must balance the friction between 2 and 1.

$$\underline{binary \rightarrow ternary}$$

binary :

$$\frac{\Delta x_1}{\bar{x}_1} = \bar{x}_2 \frac{\bar{v}_2 - \bar{v}_1}{k_{12}}$$

ternary

$$\frac{\Delta x_1}{\bar{x}_1} = \bar{x}_2 \frac{\bar{v}_2 - \bar{v}_1}{k_{12}} + \bar{x}_3 \frac{\bar{v}_3 - \bar{v}_1}{k_{13}}$$

$$\frac{\Delta x_2}{\bar{x}_2} = \bar{x}_1 \frac{\bar{v}_1 - \bar{v}_2}{k_{21}} + \bar{x}_3 \frac{\bar{v}_3 - \bar{v}_2}{k_{23}}$$

friction (1 on 2) =

$$- \text{friction (2 on 1)}$$

$$\longrightarrow \boxed{k_{12} = k_{21}}$$

Fig. 6.1

Remember that the transport relation for a component is that per mole of that component. A look at the equations then shows that the mass transfer coefficients between 1 and 2 and 2 and 1 must be equal.

A CONDENSER

Your first multicomponent exercise concerns a condenser (Fig. 6.2). The tubes of the condenser are cooled internally, and water and ammonia condense from a vapour mixture of ammonia, water and hydrogen. The hydrogen is insoluble and does not condense.

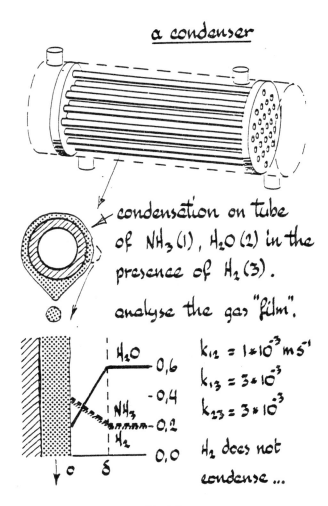

a condenser

condensation on tube of NH_3 (1), H_2O (2) in the presence of H_2 (3).

analyse the gas "film".

$k_{12} = 1 * 10^{-3} \, m s^{-1}$

$k_{13} = 3 * 10^{-3}$

$k_{23} = 3 * 10^{-3}$

H_2 does not condense ...

Fig. 6.2

It is important that you try this exercise yourself. Further on in this book there will be many more examples of multicomponent transfer (usually more complicated). Once you have the idea you will be able to glance over them and you will recognize the same things coming back again and again. But you must first try yourself. All data for the exercise are given in the figure. So go ahead. Write down the transport equations and the bootstrap relations and solve the velocities. It may cost you a quarter of an hour, but once you have done it you will see how simple it is.

The exercise is worked out in Fig. 6.3. If you write out the equations you have a set of three linear equations in the three velocities. In this example the bootstrap relation is so simple that you can immediately reduce the system to two equations and solve them.

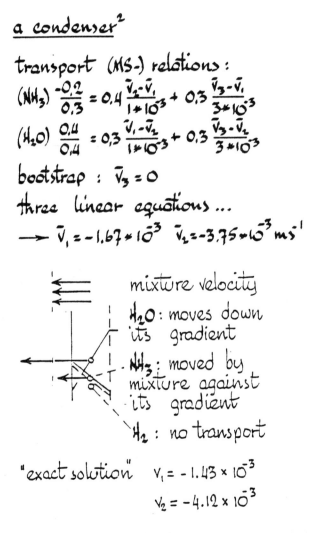

Fig. 6.3

The mixture moves towards the liquid. The driving force on water is in the same direction and it moves more rapidly. Ammonia is retarded by a driving force in the opposite direction. Hydrogen does not move at all. Its driving force cancels against friction with the other components.

If you know the driving forces, the theory here always gives a linear set of equations for the velocities. The problems arise when you do not know the driving forces. You then usually have to solve the equations by trial and error. Even then, the fact that the velocity equations are linear is a distinct advantage.

A TERNARY DISTILLATION

We now turn to another multicomponent problem: the prediction of tray efficiencies in distillation. Doing this completely would require setting up a complete model of a tray and this is outside the scope of this book. We can, however, obtain a qualitative idea of the problems by looking at our transport relations.

The (Murphree) efficiency of a tray is defined as the ratio of two quantities. The first is the change of the vapour composition on a real tray for a certain component. The second is the change that would occur in an equilibrium stage. Such stage is an ideal concept: you may regard it as a tray with both phases well mixed and no mass transfer resistances. In binary distillation, efficiencies have been found to be very useful. They are always positive and usually have a value somewhat below one. They tend to be fairly constant throughout a column, and are easily incorporated in distillation design. Can analogous concepts be used in multicomponent distillation? We are afraid not and will try to explain this.

Look at the vapour film of the distillation column given in Fig. 6.4. There are three components: ethanol, water and a trace of butanol. Ethanol is vaporizing; it has a gradient from the liquid down to the gas. Water has an opposite gradient and is condensing. The butanol compositions have been chosen equal at both sides of the film: butanol has no gradient. The transfer coefficients involving water are much higher than those involving the alcohols, and the bootstrap relation is that of 'equimolar overflow' as already discussed.

The friction between butanol and ethanol is much larger than that between butanol and water. Now the usual question: will butanol be moving, and in which direction? Of course butanol will be dragged along by the ethanol: it will be vaporizing. However, butanol has no driving force: its vapour and liquid compositions are in equilibrium. So the butanol efficiency must be indeterminate — plus or minus infinity!

We look at a few small variations of the butanol gradient (Fig.

a ternary distillation
-the vapour film -

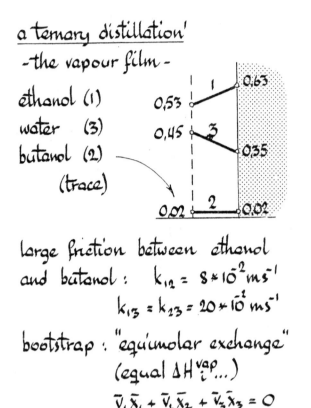

ethanol (1)   0.53

water   (3)   0.45

butanol (2)

(trace)

large friction between ethanol

and butanol : $k_{12} = 8 * 10^{-2}\, ms^{-1}$

$$k_{13} = k_{23} = 20 * 10^{-1}\, ms^{-1}$$

bootstrap : "equimolar exchange"

       (equal $\Delta H_i^{vap}...$)

$$\bar{V}_1 \bar{x}_1 + \bar{V}_2 \bar{x}_2 + \bar{V}_3 \bar{x}_3 = 0$$

in which direction does (2) move?

Fig. 6.4

6.5). A positive gradient increases the negative velocity obtained with a zero gradient. So with a small positive gradient the efficiency must be strongly positive (and much larger than one). With a very small negative gradient, butanol will still be moving up its gradient and its efficiency will be negative. Only when the gradient becomes sufficiently negative will the friction forces be overcome so that the efficiency will be positive again.

More complete calculations confirm the picture found here. Murphree efficiencies in multicomponent mixtures are not the simple well-behaved functions of binary systems. The binary concept is useless. Please note that the problems are not due to non-idealities. In this case they are due to ideal gas interactions!

We believe efficiency approaches to multicomponent distillation to be a dead-end alley. They only increase confusion. We prefer models in which the mass transfer fluxes are computed directly.

It may be remarked at this point that the 'three gases' problem

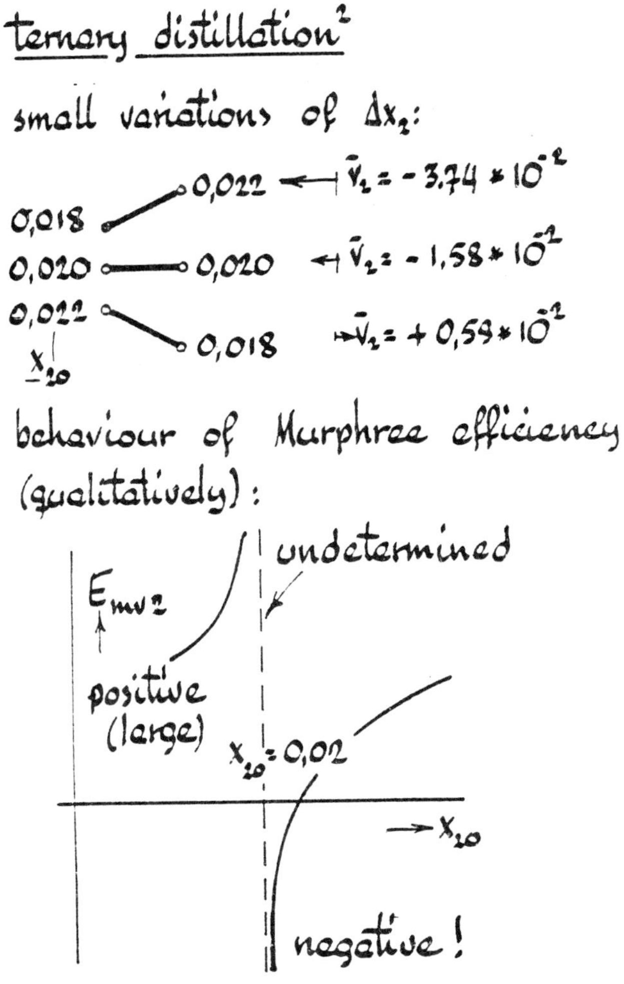
ternary distillation²
small variations of Δx₂:
0,018
0,020
0,022
x₂₀
0,022 ← v̄₂ = -3.74 * 10⁻²
0,020 ← v̄₂ = -1.58 * 10⁻²
0,018 → v̄₂ = +0,59 * 10⁻²
behaviour of Murphree efficiency
(qualitatively):
undetermined
Emv2
positive
(large)
x₂₀ = 0,02
→ x₂₀
negative!

Fig. 6.5

from Chapter 2 is formally identical to the problem we have just considered. It can be described almost quantitatively with the same equations.

BINARY APPROXIMATION OF A TERNARY

Binary mass transfer relations are often used by engineers for ternary mixtures. From the previous examples you might obtain the impression that this should lead to disaster. This is not so: in most cases the problems are less spectacular.

When can ternaries be approximated as binaries? Three cases are shown in Fig. 6.6.

The first occurs if all species are similar. The three transfer coefficients may then be (approximately) equal. A nice example might be diffusion of the three isomers of xylene. As you may easily

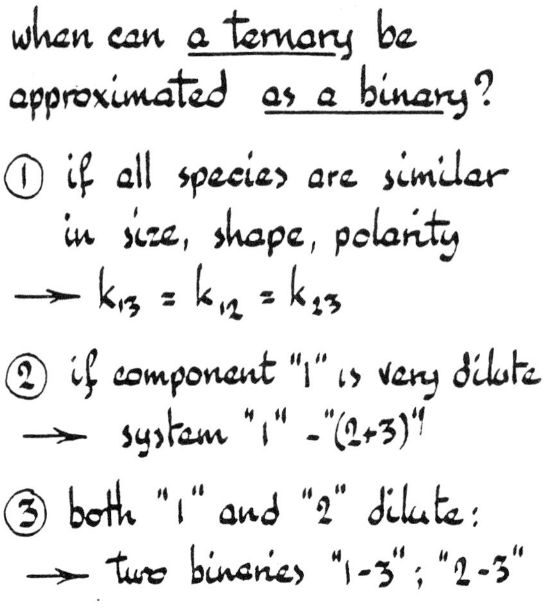

when can a ternary be
approximated as a binary?

① if all species are similar
in size, shape, polarity
$\longrightarrow k_{13} = k_{12} = k_{23}$

② if component "1" is very dilute
$\longrightarrow$ system "1" - "(2+3)"

③ both "1" and "2" dilute:
$\longrightarrow$ two binaries "1-3"; "2-3"

Fig. 6.6

check, the equations of each component can be written in such a way that the other two are grouped as one component.

The second case occurs when one component, say '1', is very dilute. '2' and '3' then behave as a binary.

In the third case, one component — the solvent — dominates. The two other components then diffuse independently.

Many problems are approximately described by these situations. Their behaviour is then similar to that of a binary. You should not, however, take this as an excuse for not learning the multicomponent approach. If you use the binary viewpoint you cannot see when multicomponent effects do become important.

SUMMARY

- In the driving force/friction description of mass transfer, multi-component systems are a straightforward extension of binary systems.
- You need one mass transfer coefficient for each pair of inter-actions. So one for a binary, three for a ternary, six for a quaternary and ten for a five-component mixture.
- For given boundary conditions of the film the transport and bootstrap relations are linear equations in the velocity. These are easily solved.
- In multicomponent systems, friction can well cause a component to move against its gradient. This is why certain binary concepts

such as that of a tray efficiency are not always transferable to multicomponent systems.

- Binary theory can be used to approximate multicomponent systems when all mass transfer coefficients are equal, and when one or two components are very dilute.

7

Mass and heat transfer

In this chapter we study the effect of temperature gradients on mass transfer.

TEMPERATURE GRADIENTS

In many mass transfer processes, such as condensation, evaporation and drying, we encounter temperature gradients within the diffusional 'film'. This is shown qualitatively in Fig. 7.1. Such gradients are

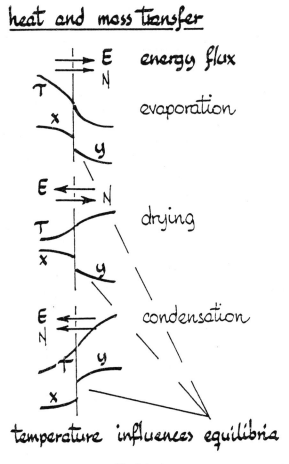

Fig. 7.1

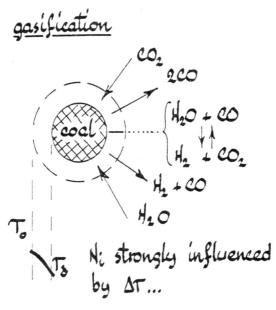

Fig. 7.2

also found in systems where chemical reactions with large heat effects occur. An example is the combustion or gasification of coal (Fig. 7.2).

Temperature gradients have several effects on mass transfer:

- they are the direct cause of molecular motion,
- they may cause bulk motion of the fluid, and
- they influence phase equilibria and reaction rates.

The first of these effects is known as thermal diffusion. It plays a role in the separation of isotopes and in thermal deposition of dust from gases. Otherwise the effect of thermal diffusion is usually very small and we will not take it into account.

Temperature alters the bulk and interfacial properties of fluids. Such variations can give rise to 'free convection' and interfacial motion (Bénard effect). Both effects can be taken into account via the mass transfer coefficients in our models.

Finally, temperature levels influence phase equilibria and reaction rates. In that way they influence mass transfer rates. Also mass transfer fluxes transfer heat and so affect temperature gradients. These simultaneous effects are the subject of this chapter.

ENTHALPY

We begin with a brief repetition of a few aspects of energy transfer. We take one mole of a substance and add heat, keeping the pressure constant (Fig. 7.3). The heat taken up by the substance is known as its 'change of enthalpy'. This enthalpy is a property of the substance. It

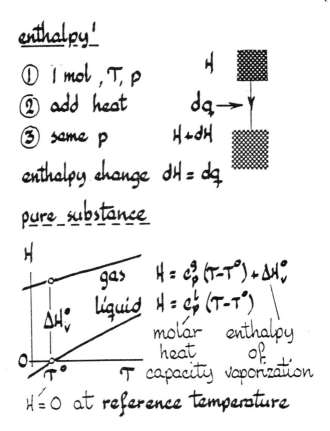

enthalpy'

① 1 mol, T, p   H

② add heat $dq \rightarrow$

③ same p   H + dH

enthalpy change dH = dq

pure substance

H

gas $H = c_p^g (T - T^o) + \Delta H_v^o$

ΔH_v^o liquid $H = c_p^l (T - T^o)$

O T^o T   molar   enthalpy
                                        heat   of
                                        capacity   vaporization

$H = 0$ at reference temperature

Fig. 7.3

depends on temperature and pressure. Except in the vicinity of the critical point, the effect of pressure is small however.

For a liquid, enthalpy increases more or less linearly with temperature. The proportionality constant is known as the molar heat capacity. In transport processes the absolute value of the enthalpy is unimportant: we only need enthalpy differences. The enthalpy can then be put equal to zero at any convenient reference temperature.

If the liquid is vaporized, the enthalpy increase is known as the 'enthalpy of vaporization'. The gas also has an enthalpy which is again a function of temperature. As the molar heats of gas and liquid usually differ, the enthalpy of vaporization does depend on the temperature. However, it does not vary rapidly.

In a mixture (Fig. 7.4), the enthalpy is a summation of contributions from the different species. These 'partial molar enthalpies' depend on composition. In the examples in this chapter, we will neglect such variations.

If chemical reactions occur between species, we cannot assign independent zero points to the enthalpies of all components. As an example we look at the nitrogen–hydrogen–ammonia system (Fig. 7.4). We assign a value of zero to both reactants at the same

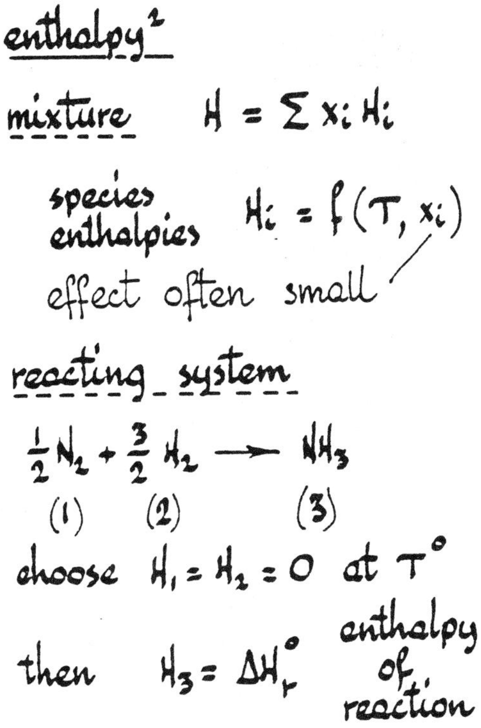

$$\underline{enthalpy^2}$$

$$\underline{mixture} \quad H = \sum x_i H_i$$

$$\begin{array}{l} species \\ enthalpies \end{array} \quad H_i = f(T, x_i)$$

$$effect \ often \ small$$

$$\underline{reacting \ system}$$

$$\tfrac{1}{2} N_2 + \tfrac{3}{2} H_2 \longrightarrow NH_3$$

$$(1) \qquad (2) \qquad \qquad (3)$$

$$choose \ H_1 = H_2 = 0 \ at \ T^\circ$$

$$then \quad H_3 = \Delta H_r^\circ \quad \begin{array}{l} enthalpy \\ of \\ reaction \end{array}$$

Fig. 7.4

temperature. Then the enthalpy of ammonia is equal to the 'enthalpy of reaction' at that temperature.

The energy flux through a film is given in Fig. 7.5. There are two contributions:

- heat conduction and
- enthalpy carried by the mass transfer fluxes.

We approximate this relation by a difference equation involving a 'heat transfer coefficient'.

CONDENSATION IN PRESENCE OF AN INERT GAS

To illustrate combined heat and mass transfer we consider condensation of methanol in the presence of nitrogen. Nitrogen does not condense (Fig. 7.6). The bulk vapour composition and temperature are given, and also the bulk coolant temperature. The questions are:

- what is the flux of methanol and
- what is the energy flux?

Here we only analyse the gas film. The relation between the energy flux in the liquid and the interfacial temperature is taken as that given in the figure.

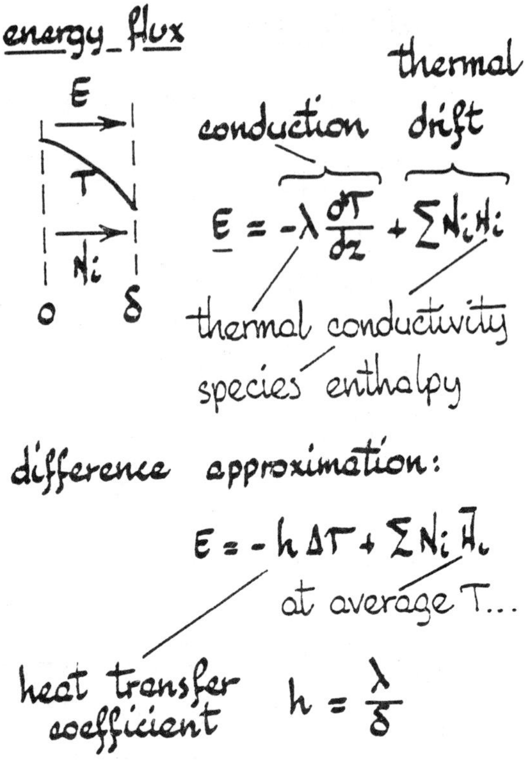

energy flux   thermal

E

conduction drift

T

$$E = -\lambda \frac{dT}{dz} + \sum N_i H_i$$

N_i

$0 \quad \delta$

thermal conductivity

species enthalpy

difference approximation:

$$E = -h \Delta T + \sum N_i \overline{H_i}$$

at average T...

heat transfer
coefficient $h = \frac{\lambda}{\delta}$

Fig. 7.5

Let us have a look at the interface and assume that we know its temperature. Also we assume that vapour and liquid are in equilibrium at the interface. The vapour pressure relation of methanol then gives the interfacial composition of the vapour. We can now calculate the methanol flux in the same way as in Chapter 5.

Calculating the energy fluxes to and from the interface is now straightforward. These two fluxes should of course be equal. Otherwise we try a new temperature until they are (Fig. 7.7). You could also have solved all equations in one go with your equation solver. This is the preferred way in more complicated situations.

HETEROGENEOUS REACTING SYSTEMS

As a second example we have a look at the synthesis of ammonia (Fig. 7.8). The reaction is exothermic, so heat is transferred from the particle to the gas. The situation is typical for that in many catalytic reactors. The energy transport relations should be obvious. In this case there is a net flow towards the interface. This acts against the temperature gradient. There is a negative thermal drift for the ammonia synthesis. If we ignore thermal drift in our calculation we obtain an error in the temperature estimation. Since the surface

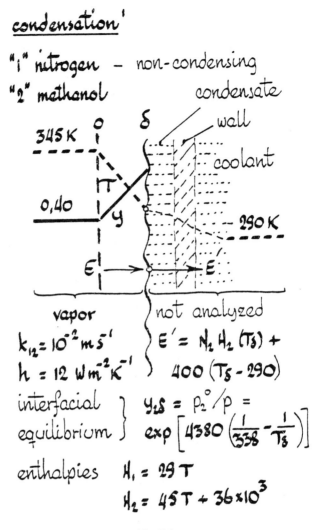

condensation'

"1" nitrogen – non-condensing
"2" methanol   condensate

345 K

0,40

vapor

$k_{12} = 10^{-2} m s^{-1}$

$h = 12 W m^{-2} K^{-1}$

interfacial
equilibrium

enthalpies

not analyzed

$E' = N_2 H_2 (T_\delta) + 400 (T_\delta - 290)$

$y_{2\delta} = P_2^o / P = \exp \left[4380 \left(\frac{1}{338} - \frac{1}{T_\delta} \right) \right]$

$H_1 = 29 T$

$H_2 = 45 T + 36 \times 10^3$

Fig. 7.6

reaction rate is a strong function of temperature, this error may be amplified in reactor design calculations. Standard texts on chemical reaction engineering usually ignore thermal drift; this is not always justified.

AN AMMONIA ABSORBER

Finally we consider absorption of ammonia from air into water in a packed column (Fig. 7.9). This problem is not as simple as it looks.

We assume counter-current operation, with fresh water entering at the top. The rich ammonia/air mixture enters at the bottom where the ammonia is absorbed. The enthalpy of vaporization is released; this causes a rise in temperature of the liquid. As a result, water

condensation[2]

① assume values for T_s

② calculate y_s (T_s)

③ calculate N_2 (T_s)

(as in previous chapters)

④ calculate

$$E(T_s) = h(T_o - T_s) + N_2 H_2(\bar{T})$$

$$E'(T_s) = N_2 H_2(T_s) + 400(T_s - 290)$$

⑤ at the actual value of T_s

these should be equal:

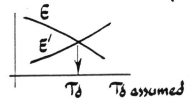

Fig. 7.7

vaporizes. The mass transfer process in the vapour therefore involves three species: ammonia, water and (stagnant) air. Ammonia and water vapour diffuse against each other at the bottom of the column.

Towards the top of the column the vapour encounters cold entering water. Therefore water vapour **condenses** near the top of the column and we now have co-diffusion of ammonia and water through air. Water vaporization at the bottom and condensation at the top cannot be ignored in the analysis. The resulting temperature profiles along the column show a pronounced bulge towards the bottom.

SUMMARY

- Temperature gradients influence mass transfer in several ways. Most important, they influence phase equilibria and chemical reaction rates. In that way they influence the driving force for mass transfer.
- Mass fluxes in turn carry enthalpy fluxes with them. The contribution of this thermal drift is significant in several applications such as condensation and heterogeneous catalytic reactors.

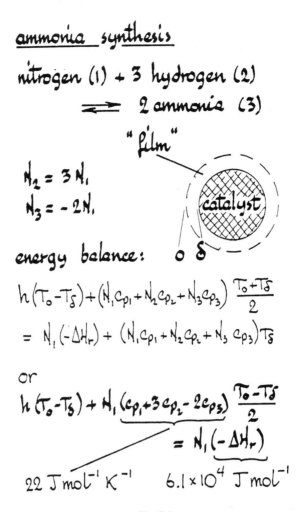

ammonia synthesis

nitrogen (1) + 3 hydrogen (2)

$\rightleftharpoons$ 2 ammonia (3)

"film"

$N_2 = 3 N_1$

$N_3 = -2 N_1$

energy balance: o δ

$$h(T_o - T_\delta) + (N_1 c_{p_1} + N_2 c_{p_2} + N_3 c_{p_3}) \frac{T_o + T_\delta}{2}$$

$$= N_1(-\Delta H_r) + (N_1 c_{p_1} + N_2 c_{p_2} + N_3 c_{p_3}) T_\delta$$

or

$$h(T_o - T_\delta) + N_1 \underbrace{(c_{p_1} + 3 c_{p_2} - 2 c_{p_3})}_{22 \ J \ mol^{-1} \ K^{-1}} \frac{T_o - T_\delta}{2}$$

$$= N_1 \underbrace{(-\Delta H_r)}_{6.1 \times 10^4 \ J \ mol^{-1}}$$

Fig. 7.8

- If energy and mass transfer occur together, you must solve the energy and transport relations simultaneously. The energy relation to remember is the difference equation in Fig. 7.5.

absorption

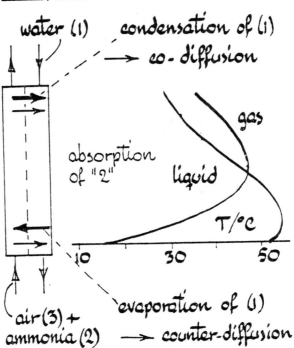

Fig. 7.9

8

Non-idealities

So far we have only considered ideal solutions (mainly gases). This was just to keep our equations simple and tidy: in our new mass transfer theory it is not difficult to take the effects of non-ideality into account. Non-idealities only affect the composition term in the driving force (Fig. 8.1).

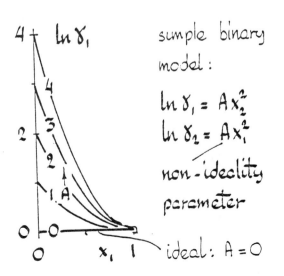

non-'ideality'

$$\frac{\Delta \mu_i}{RT} = \Delta \ln(\gamma x)_i = \Delta \ln \gamma_i + \Delta \ln x_i$$

affects the composition term in the "driving force".

simple binary model:

$$\ln \gamma_1 = A x_2^2$$
$$\ln \gamma_2 = A x_1^2$$

non-ideality parameter

ideal: $A = 0$

Fig. 8.1

A SIMPLE MODEL OF NON-IDEALITIES

To take non-ideality into account you must be able to predict activity coefficients. For gas–liquid systems (and to a reasonable extent also for liquid–liquid and liquid–solid systems) thermodynamic models are available to do this. Well-known ones are the Van Laar, Wilson, NRTL and UNIQUAC models. To give a reasonably accurate description of real systems, the functions used in these models have to be fairly complicated. However, their details are not of importance for understanding mass transfer in non-ideal systems. So the greater part of this discussion will be done using the simplest model that we can think of. (See the bottom part of Fig. 8.1.) This is a model for a binary system. The natural logarithm of an activity coefficient is equal to a constant A times the square of the mole fraction of the other component. This parameter A is a measure of the non-ideality of the system. If A is zero, the system is ideal. Increasingly positive values of A correspond to increasing non-idealities.

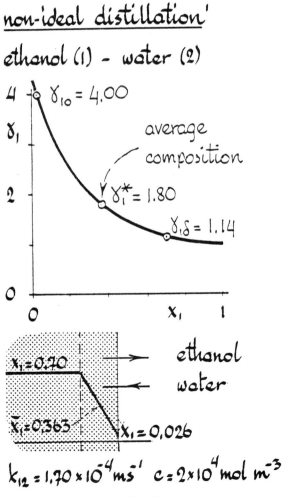

Fig. 8.2

NON-IDEAL BINARY DISTILLATION

As an example we look at the liquid phase in the distillation of water and ethanol (Fig. 8.2).

The non-ideality of ethanol–water can be described roughly by our simple model using a value of A of 1.2. (We have, however, plotted accurate values and used these in the calculations.) The ethanol fractions at the two sides of the film are given in the lower figure. So is the mass transfer coefficient and the total concentration. In this example the total concentration varies considerably in the film. It has been evaluated at the average concentration in the film.

We have worked out the driving force including non-idealities for ethanol in Fig. 8.3. There is one point to note in the difference approximation used. In the nominator you should use the activity coefficient at the average composition. This is considerably more accurate than using the average activity coefficient.

You see that it is all quite straightforward. In this example the driving force chosen is very large. (You will seldom encounter such a large composition difference in the liquid of a distillation tray; this is mainly because mass transfer in distillation is usually limited by the

$$\underline{\text{non-ideal distillation}^2}$$

$$\text{bootstrap}: \text{"equimolar exchange"}$$

$$\bar{V}_1 \bar{X}_1 = -\bar{V}_2 \bar{X}_2$$

$$\text{transport}:$$

$$\frac{\Delta(\gamma_1 x_1)}{(\gamma_1 x_1)^*} = \bar{X}_2 \frac{\bar{V}_2 - \bar{V}_1}{k_{12}}$$

$$\text{at the average composition}$$

$$\frac{4.00 \times 0.026 - 1.14 \times 0.70}{1.80 \times 0.363} = -1.06$$

$$(-1.86 \text{ if non-ideality is neglected})$$

$$\text{working out yields:}$$

$$V_1 = 1.7 \times 10^{-4} \times 1.06 = 1.80 \times 10^{-4} \, ms^{-1}$$

$$N_1 = V_1 c \bar{X}_1 = 1.31 \, mol \, m^{-2} s^{-1}$$

$$(\text{the "exact" value is } 1.13)$$

Fig. 8.3

resistance in the vapour phase.) Also the non-ideality is considerable. Even so, the difference approximation is not far from the 'exact' result, where varying concentrations and activity coefficients in the film have been taken into account.

LARGE NON-IDEALITIES: DEMIXING

Back again to our simple model of non-idealities. We have plotted the dimensionless potential of a component against its mole fraction in Fig. 8.4. With the logarithmic composition scale used, this gives a

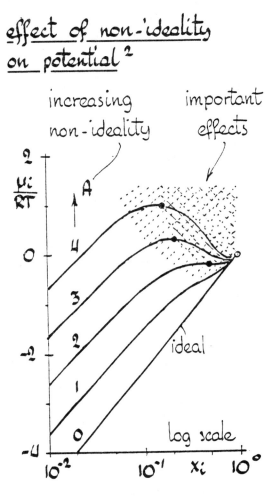

Fig. 8.4

straight line for an ideal solution. Also for dilute non-ideal systems, straight lines are observed. As far as mass transfer goes, these can be regarded as ideal. (This is also the case when a component mole fraction goes to unity. However this is not so obvious with the plotting used in Fig. 8.4).

In the upper right-hand side of the figure a number of interesting things can be observed. For $A=1$ (a moderately non-ideal system) the potential still rises monotonically with composition. For $A=2$ the curve has one point with a horizontal tangent (at $x=0.5$). This is a 'critical point'. We shall see in a moment what this means.

For $A>2$ there is a domain where the component potential decreases with increasing mole fraction (see also Fig. 8.5). This domain is bounded by the two 'spinodal points'. Within the spinodal domain the mixture is completely unstable: a component tends to diffuse to regions where it has a higher concentration. The mixture then splits in two parts: one with a high and one with a low mole fraction of that component.

A complete analysis of what happens is outside the scope of this book. We have, however, summarized the main results at the bottom of Fig. 8.5. For $A<2$ there is no demixing. The critical point ($x=0.5$; $A=2$) is at the boundary of the demixing zone. For $A>2$ we see the

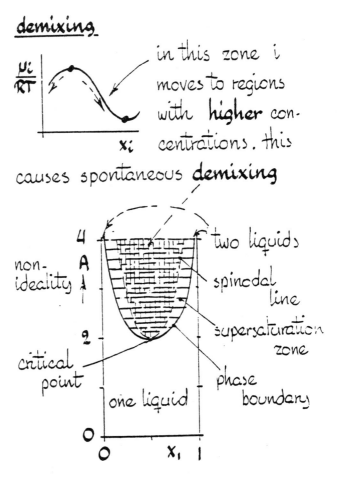

Fig. 8.5

spinodal zone where demixing occurs spontaneously. It is bounded by two supersaturation zones. Fluids with a composition in such a zone can exist indefinitely, but they are metastable. Any disturbance tends to let them split into two liquids with compositions on the 'phase boundary'.

MAXWELL–STEFAN VERSUS FICK

This book uses the Maxwell–Stefan relations to describe mass transfer. Most other texts use the classical Fick description, and most diffusivity data you will encounter in literature are given as Fick coefficients. So it is important to have some idea of the relations between the two kinds of coefficients. In the conventional theory the effects of non-ideality are incorporated in the Fick diffusivity. In the generalized Maxwell–Stefan (MS) approach they are part of the driving force (Fig. 8.6).

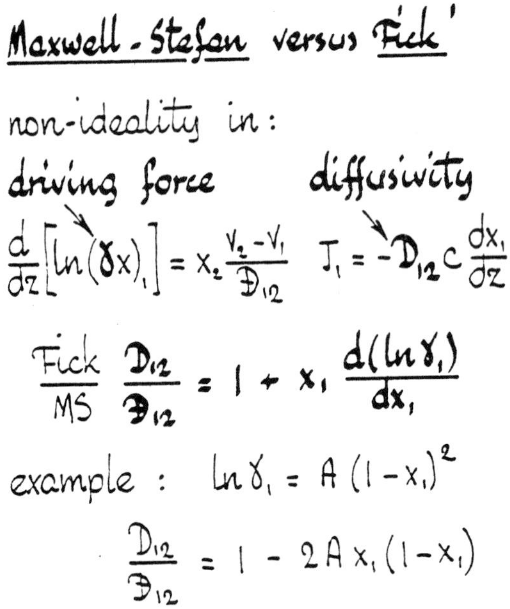

$$\underline{\text{Maxwell - Stefan}} \text{ versus } \underline{\text{'Fick'}}$$

non-ideality in:

driving force   diffusivity

$$\frac{d}{dz}\left[\ln(\gamma x)_1\right] = x_2 \frac{v_2 - v_1}{\mathcal{D}_{12}} \qquad J_1 = -D_{12}\, c\, \frac{dx_1}{dz}$$

$$\frac{\text{Fick}}{\text{MS}} \quad \frac{D_{12}}{\mathcal{D}_{12}} = 1 + x_1 \frac{d(\ln\gamma_1)}{dx_1}$$

example : $\ln\gamma_1 = A\,(1-x_1)^2$

$$\frac{D_{12}}{\mathcal{D}_{12}} = 1 - 2A\,x_1\,(1-x_1)$$

Fig. 8.6

For a binary mixture the relation between Fick and MS diffusivities is given in the figure. The relation is not difficult to derive: you need to replace the velocities in the MS equation by fluxes with respect to the mixture. For our simple model the ratio of the two diffusivities is given at the bottom of the figure.

The behaviour of the Fick diffusivities is summarized in Fig. 8.7. Before having a look at it you should know that the MS diffusivity is always a positive, usually monotonic and well-behaved function.

For an ideal binary solution the Fick and MS diffusivities are

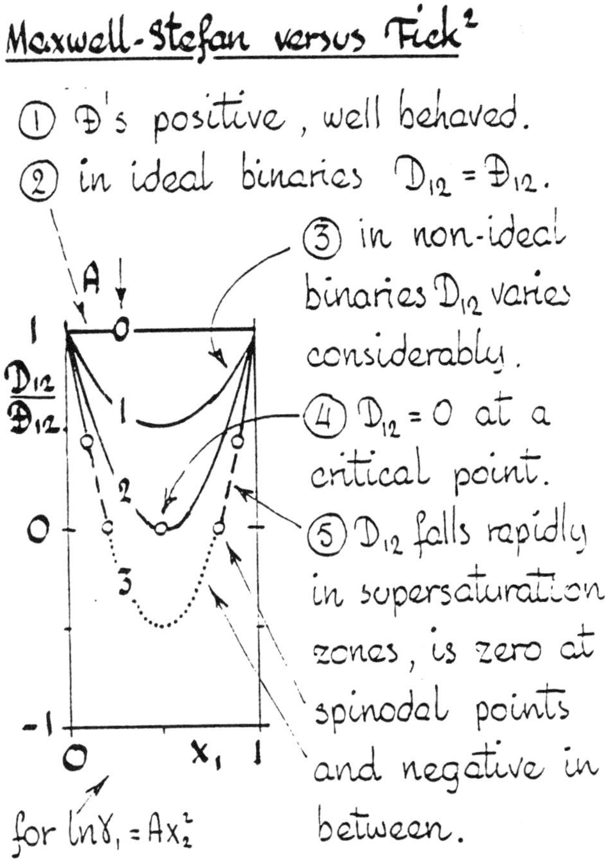

Maxwell-Stefan versus Fick?

① $\mathcal{D}$'s positive, well behaved.

② in ideal binaries $\mathcal{D}_{12} = Ð_{12}$.

③ in non-ideal binaries $\mathcal{D}_{12}$ varies considerably.

④ $\mathcal{D}_{12} = 0$ at a critical point.

⑤ $\mathcal{D}_{12}$ falls rapidly in supersaturation zones, is zero at spinodal points and negative in between.

for $\ln \gamma_1 = A x_2^2$

Fig. 8.7

equal. In non-ideal solutions the Fick coefficient shows a minimum at intermediate concentrations. This minimum becomes progressively deeper as the non-ideality increases. At the critical point the Fick diffusivity becomes zero. A further increase in the non-ideality gives a spinodal zone where the Fick coefficients are negative and super-saturation zones where they rapidly fall to zero.

Clearly the behaviour of Fick diffusivities is considerably more complicated than that of MS diffusivities. This is even more so in multicomponent mixtures.

LIQUID–LIQUID EXTRACTION

Several separation processes make use of strong non-idealities in liquids. Well-known examples are liquid–liquid extraction and extractive and azeotropic distillation. In these processes there are at least three components. It is outside the scope of this book to treat them properly, but we will discuss a few aspects of their mass transfer characteristics.

The example used is a liquid–liquid system which extracts water from acetone, using glycerol as a solvent. The composition diagram of the system is shown in Fig. 8.8. Acetone and glycerol are poorly miscible; they have a demixing zone along the bottom line of the triangle. Water is miscible with both acetone and glycerol. For water concentrations that are not too high there are two phases: one rich in acetone and one rich in glycerol. Points on the phase boundary which are in equilibrium are connected by tie lines. These slope upwards to the right, showing that water has a preference for the glycerol phase.

In computer calculations, information on the activity coefficients of the components is stored in a thermodynamic model. This model can also calculate the phase boundaries and tie lines in the diagram.

The extraction is done in a series of counter-current stages. Each of these consists of a mixer for mass transfer and a settler for

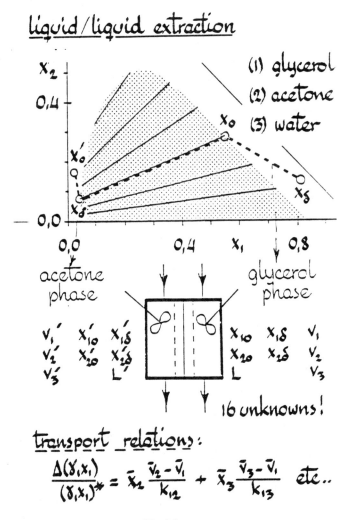

Fig. 8.8

separating the phases. We will only consider a single stage and neglect any mass transfer in the settler. In the mixer, both phases are considered to be well mixed. There is a mass transfer resistance on either side of the interface. In such strongly non-ideal systems, no simple bootstrap relations can be derived beforehand; all mass balance and transport relations have to be solved simultaneously. The variables in this model are shown in the figure. There are sixteen unknowns! (You can reduce the number of variables by three by using fluxes instead of velocities.) This complexity, by the way, has nothing to do with our using the MS equations.

The sixteen equations can be chosen as follows:

- three overall mass balances for the three components,
- three equilibrium relations at the interface,
- two transport relations in the glycerol film,
- two transport relations in the acetone film,
- three partial mass balances relating the fluxes through the interface to the flow changes in one of the phases, and
- three equality relations for the fluxes in the two films.

As you can see, the bootstrapping of this problem is not a trivial task.

On the other hand, the transport relations of each film separately are nothing special. We have written one of them down for the glycerol phase. It is a simple extension of the binary case; hardly anything to be remarked upon.

SUMMARY

- For non-ideal mixtures you need a thermodynamic model which predicts activity coefficients. If you have such a model, including the non-ideality effects in our mass transfer theory is trivial. Just have a look back at Fig. 8.3 to see how that is done.
- Dilute systems behave as ideal systems.
- Non-idealities become particularly important in the vicinity of a critical point. There is, however, nothing special in the behaviour of the MS diffusivities around such a point. The same is true in the supersaturation zones. This is because the effect of non-idealities is included in the driving forces.
- In contrast with the MS diffusivities the Fick diffusivities of classic mass transfer do show a complicated behaviour in non-ideal mixtures.
- In liquid–liquid extraction (a fairly typical non-ideal process) bootstrapping requires the simultaneous solution of many equations. The transport relations, however, are nothing special.

9

Transport coefficients

In our transport relations we need multicomponent mass transfer coefficients. These are calculated from Maxwell–Stefan diffusivities. This chapter summarizes our knowledge in this area for gases and non-electrolyte liquids.

DIFFUSIVITIES IN GASES

Gas diffusivities can be estimated quite well from the kinetic theory of gases. In the simplest form of this theory, molecules are assumed to be hard spheres. Friction between species is caused by momentum transfer in binary collisions of unlike molecules. The derivation is not extremely difficult, but too long to give here. Important results (Fig. 9.1) are:

- friction is proportional to the velocity difference between the species,
- the diffusion coefficients do not depend on the gas composition,
- diffusivities increase rapidly with temperature and are inversely proportional to the pressure,
- diffusivities are lower for larger and heavier molecules.

Real molecules are not hard spheres. At higher temperatures they undergo harder collisions and then show a lower effective diameter. This is taken into account in the empirical modification of the theory at the bottom of the figure. To use this equation you need molar 'diffusion volumes'. These are tabulated in handbooks; their values are roughly two thirds of the corresponding liquid volumes (Fig. 9.2). The equation is easy to use. Errors are less than 10% up to pressures of 1 MPa. At higher pressures the assumptions of the kinetic theory:

- only binary collisions and
- a 'free path' much larger than the molecule diameters

are not correct.

kinetic gas theory.

hard sphere model $\;d_1$

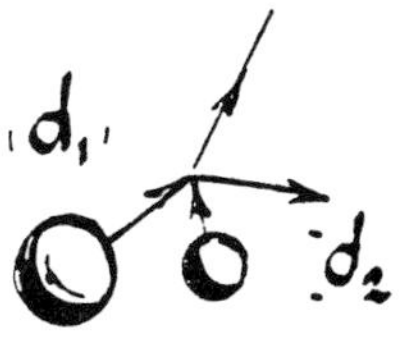

$$D_{12} = \frac{2^{1/2}}{\pi^{3/2}} \frac{(RT)^{3/2}}{L\rho d_{12}^2} \left(\frac{1}{M_1} + \frac{1}{M_2}\right)^{1/2}$$

$$d_{12} = (d_1 + d_2)/2$$

empirical modification:

$$D_{12} = 3.16 \times 10^{-8} \frac{T^{1.75}}{p(\sigma_1^{1/3} + \sigma_2^{1/3})^2} \left(\frac{1}{M_1} + \frac{1}{M_2}\right)^{1/2}$$

diffusion volume, $m^3 mol^{-1}$
($\lesssim$ liquid volume)

Fig. 9.1

kinetic gas theory ^{2}

H_2	N_2	CO_2	NH_3	H_2O
7.07	17.9	26.9	14.9	12.7

σ_i / $10^{-6} m^3 mol^{-1}$

example: $N_2 (1) \quad CO_2 (2)$

$\qquad T = 300\ K \quad p = 10^5 Pa$

$$D_{12} = 3.16 \times 10^{-8} \frac{300^{1.75}}{10^5 ((17.9 \times 10^{-6})^{1/3} + (26.9 \times 10^{-6})^{1/3})^2}$$

$$\times \left(\frac{10^3}{28} + \frac{10^3}{44}\right)^{1/2} = 1.75 \times 10^{-5} m^2 s^{-1}$$

Fig. 9.2

The friction is determined by binary collisions. For mixtures of more than two components you can therefore simply use the binary coefficients (Fig. 9.3). This is a distinct advantage of the Maxwell–Stefan approach. It is, however, only applicable to ideal gases.

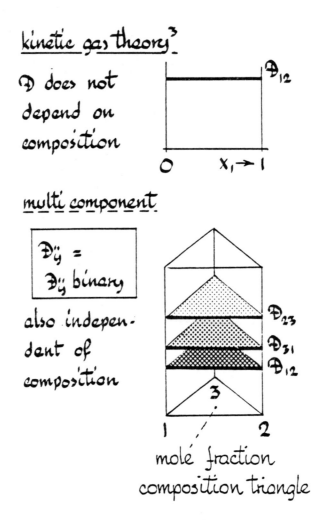

Fig. 9.3

DIFFUSIVITIES IN LIQUIDS

For liquids the situation is less satisfactory. There is no good, accurate theory. The 'sphere in liquid' theory that we have seen earlier gives the right orders of magnitude for dilute binary solutions (Fig. 9.4). For mixtures of small molecules it can be improved by

liquids - dilute solutions

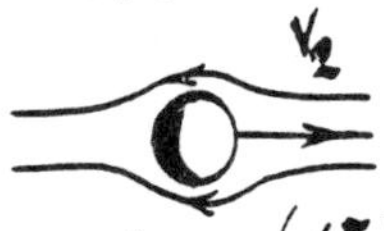

in pure "1"

$$d_2 = \left(\frac{v_2}{L}\right)^{1/3}$$

$$D_{12}^{x_1=1} = \frac{kT}{3\pi \eta_1 d_2}$$

for small molecules $d_2 \lesssim d_1$

2^- is better $\left(\frac{k}{2\pi} = 2.20 * 10^{-24}\right)$

solvent	solute	$\dfrac{T}{K}$	$\dfrac{D_{12}^{x_1=1}}{10^{-9} m^2 s^{-1}}$	$\dfrac{D_{12}^{x_1=1} 3\pi \eta_1 d_2}{kT}$
water	CO_2	298	1,92	1,50
benzene	CCl_4	298	1,92	1,51
benzene	benzene*	373	6,15	1,59
ethanol	ethanol*	280	0,62	1,08
ethanol	CO_2	298	3,20	3,04 (!)

* isotope

Fig. 9.4

changing the theoretical constant from '3' to '2'. Experimental and calculated diffusivities then usually differ by less than a factor of two. There have been many attempts to improve upon this, of which the empirical relation of Wilke and Chang is the most well known. None of these relations is reliable, however: it is clear that we still have a lot to learn about liquid diffusion.

If the binary diffusivities at infinite dilution are known, values at intermediate compositions can be estimated by logarithmic inter-polation (Fig. 9.5). This works well in about three out of four cases. In the figure we have shown two cases where it works well and one where it does not. Again there have been attempts to improve upon this, but the more complicated expressions we have seen so far do not appear to be more reliable.

In multicomponent liquids the situation is again worse. Experimental data exist only for a small number of systems. Also they usually only encompass a few measurements. For lack of something

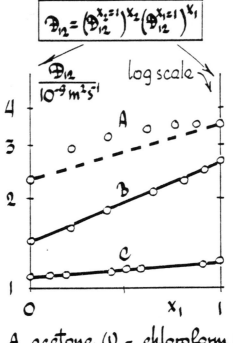

Fig. 9.5

better you might use the pseudo-binary logarithmic interpolation of Fig. 9.6. It only uses binary data. We can say one thing in its favour. The friction terms counting most heavily are those involving the highest mole fractions. These are the terms in which the compositions are relatively close to the corresponding binary.

MASS TRANSFER COEFFICIENTS

To go from diffusivities to mass transfer coefficients you need a mass transfer model. Three common models are depicted in Fig. 9.7. We already know the film theory. The resistance to mass transfer is thought to be located in a stagnant film with a certain thickness. For a membrane the thickness is a measurable quantity. In other cases it is an empirical parameter. As we have seen the multicomponent mass transfer coefficients are equal to multicomponent diffusivities divided by the film thickness.

The film theory is rather artificial, and process engineers often

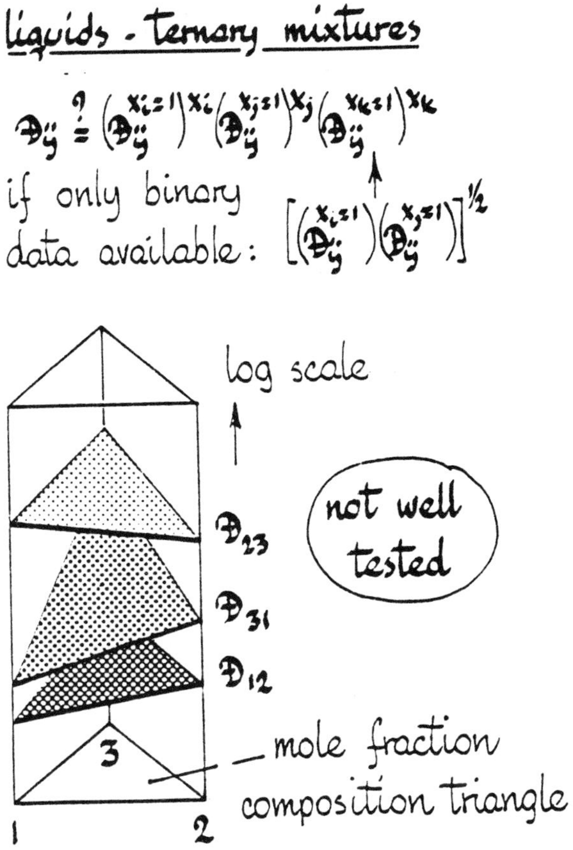

<u>liquids - ternary mixtures</u>

$$\mathcal{D}_{ij} \stackrel{?}{=} \left(\mathcal{D}_{ij}^{x_i=1}\right)^{x_i}\left(\mathcal{D}_{ij}^{x_j=1}\right)^{x_j}\left(\mathcal{D}_{ij}^{x_k=1}\right)^{x_k}$$

if only binary data available: $\left[\left(\mathcal{D}_{ij}^{x_i=1}\right)\left(\mathcal{D}_{ij}^{x_j=1}\right)\right]^{1/2}$

Fig. 9.6

prefer other models. One of these is the penetration theory. It is applicable to drops and bubbles with mobile interfaces. Here, the binary mass transfer coefficient is proportional to the square root of the diffusivity over a contact time. This time is an empirical parameter in the theory. For freely rising bubbles and drops it is approximately equal to the ratio of the diameter to the increase in velocity. To obtain multicomponent mass transfer coefficients you just plug in the appropriate multicomponent diffusion coefficients.

The same procedure can be applied with other correlations and theories that you might find in any handbook on mass transfer. For example, with solid particles you might wish to use an empirical correlation based on the boundary layer theory. Again you just use the multicomponent diffusivities in the binary expressions.

SUMMARY

- For ideal gases, binary diffusivities can be obtained from the kinetic theory of gases. Multicomponent gas diffusivities in the Maxwell–Stefan theory are identical to these binary diffusivities.

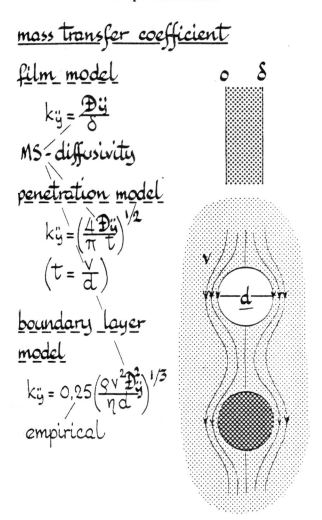

$$\underline{\text{mass transfer coefficient}}$$

$$\underline{\text{film model}} \qquad o \quad \delta$$

$$k_{ij} = \frac{\mathcal{D}_{ij}}{\delta}$$

MS - diffusivity

$$\underline{\text{penetration model}}$$

$$k_{ij} = \left(\frac{4\mathcal{D}_{ij}}{\pi\, t}\right)^{1/2}$$

$$\left(t = \frac{v}{d}\right)$$

$$\underline{\text{boundary layer}}$$
$$\underline{\text{model}}$$

$$k_{ij} = 0{,}25\left(\frac{\varrho\, v^2 \mathcal{D}_{ij}^2}{\eta\, d}\right)^{1/3}$$

empirical

Fig. 9.7

- For liquids, try at least to get experimental data for dilute solutions. If you cannot get anything, use the modified 'sphere in liquid' model or one of the empirical relations in the literature.
- For non-dilute binary solutions, use dilute diffusivities and logarithmic interpolation.
- For non-dilute multicomponent mixtures you have little choice but to use pseudo-binary logarithmic interpolation. The method is not well tested, but there is nothing better.
- Once you have the multicomponent diffusion coefficients, use them in any appropriate binary correlation or theory to obtain the multicomponent mass transfer coefficients.
- Except for gases, our knowledge of multicomponent diffusivities and mass transfer coefficients is still very incomplete.

10

Other forces

So far, the only driving forces we have considered are those due to activity gradients. There are, however, other driving forces. We consider a few of the more important ones in this chapter. These new forces give rise to extra terms in the left hand side of our transport equations.

VOLUMETRIC PROPERTIES

In the investigation of other forces we shall need two new properties of a species in a mixture:

- the species volume and
- the species density.

These properties are not necessarily equal to those of the pure species (although the differences are often not great).

The species volume (usually called the 'partial molar volume') can be determined by adding one mole of the pure species to a large amount of mixture. The resulting increase in volume (Fig. 10.1) is the species volume. The species density is simply the molar mass of the species divided by its volume.

In some cases the species properties are identical to those of the pure species. This is so:

- in ideal gases,
- in mixtures of similar molecules forming ideal solutions and
- for large solid particles such as in the following example.

PRESSURE GRADIENTS

Fig. 10.2 shows particles settling in a fluid. In this example the upward direction is taken as positive. The particles are driven downwards by

volumetric properties

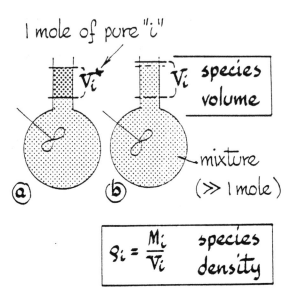

I mole of pure "i"

V_i $\quad$ V_i species volume

mixture ($\gg$ I mole)

ⓐ $\quad$ ⓑ

$$\rho_i = \frac{M_i}{V_i} \quad \text{species density}$$

Fig. 10.1

gravity. The driving force not only contains the density of the particles, but also that of the fluid. This is because the downward movement of the particles is coupled to an upward movement of an equal volume of fluid.

Gravity not only works on the particles, but also on the fluid. And as each layer of mixture weighs down on the lower layers, it creates a pressure gradient.† Here, the volume fraction of particles is very low and the pressure gradient is simply equal to the negative of the product of the fluid density and the gravitational acceleration. More generally, the pressure gradient can be calculated from the sum of all forces working on all species. (With the exception of the pressure forces that we are calculating, of course.) In this calculation the forces due to activity gradients can be omitted; they always cancel.‡ This also applies to the effect of electrical forces on ions in free solution; the positive and negative charge effects cancel. (More about this in the next chapter.)

With the pressure gradient you can eliminate the fluid properties to give a driving force which only contains species properties. There are now two forces:

---

† Here we are assuming that the pressure can be calculated from equilibrium considerations. In other words: that there is 'mechanical equilibrium' in the system. Mass transfer processes tend to be very slow compared with mechanical processes. So this is usually a very good approximation.

‡ This is known as the Gibbs–Duhem relation.

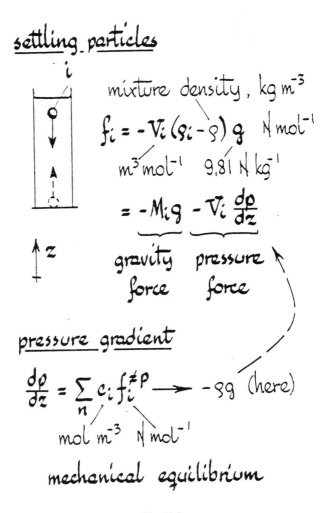

settling particles

mixture density, kg m^{-3}

$$f_i = -\overline{V}_i (g_i - g) \, g \quad \text{N mol}^{-1}$$

$$m^3 mol^{-1} \quad 9.81 \, \text{N kg}^{-1}$$

$$= -M_i g - \overline{V}_i \frac{dp}{dz}$$

$\underbrace{\quad\quad}$ $\underbrace{\quad\quad}$

gravity   pressure
force   force

pressure gradient

$$\frac{dp}{dz} = \sum_n c_i f_i^{\neq P} \longrightarrow -g g \quad (\text{here})$$

$$mol \, m^{-3} \quad \text{N mol}^{-1}$$

mechanical equilibrium

Fig. 10.2

- a gravity force, proportional to the mass of the particles and
- a pressure force proportional to their volume.

This description is often handier than the first one. However, it does not show so clearly that fields such as gravity only produce an effect if a species has a density different from the surroundings.

CENTRIFUGES

The forces due to gravity are quite insignificant in mass transfer operations. Forces which are more than five orders of magnitude larger can be obtained in centrifugal devices. In its simplest form a centrifuge is a cylindrical container, rotating rapidly around its axis. A fluid in the centrifuge is flung outward against the wall. Species denser than the fluid then move through the fluid and collect at the wall.

The centrifugal force causing this movement is similar to gravity (Fig. 10.3). There are differences: the system has a cylindrical symmetry and the force depends on the radial position. Here, we neglect these differences and assume constant average driving forces. This is allowable for shallow liquid layers.

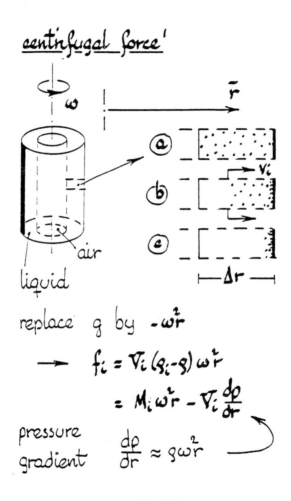

Fig. 10.3

The effect of the centrifugal field can now be obtained by a simple substitution in the gravity equation. Also this field causes a pressure gradient. If you wish you can again split the net force into contributions of the field on the species alone and a contribution from the pressure gradient. From these (and other forces) you can calculate the radial species velocities in the machine.

A difference approximation for the centrifugal force is given in Fig. 10.4. This is also only applicable to shallow liquid layers. Notice the form of the pressure term: we will be coming back to the pressure field a little further on.

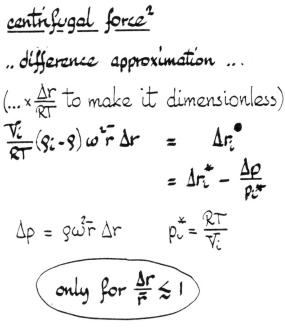

centrifugal force²

.. difference approximation ...

$(... \times \frac{\Delta r}{RT}$ to make it dimensionless$)$

$$\frac{V_i}{RT}(g_i - g)\omega^2 r \Delta r \;=\; \Delta r_i^{\bullet}$$

$$= \Delta r_i^* - \frac{\Delta p}{p_i^*}$$

$$\Delta p = g\omega^2 r \Delta r \qquad p_i^* = \frac{RT}{V_i}$$

only for $\frac{\Delta r}{r} \lesssim 1$

Fig. 10.4

The pressure difference over a centrifuge can be over one hundred megapascals! The equipment has to be very sturdy to withstand this.

PROTEIN CENTRIFUGATION

This is just one example of the application of the theory of this chapter. Centrifugal forces can be large enough to concentrate molecular species. For small molecules, only a modest concentration difference is obtained (Fig. 10.5). Larger molecules can be strongly concentrated, however. Notice that you need to use different values for the radii in the two cases.

One application is in the concentration of proteins (macro-molecules with a diameter of several nanometres) in the laboratory. However, even after a prolonged centrifugation, no sharp separation is obtained. The protein concentrates in a diffuse layer at the rim of the centrifuge. This situation is a dynamic equilibrium in which three forces on the protein balance:

- the centrifugal force outward, and
- the pressure and composition forces inward.

The thickness of the 'fuzzy' layer is easily estimated with the difference approximation. As you see it is not negligible.

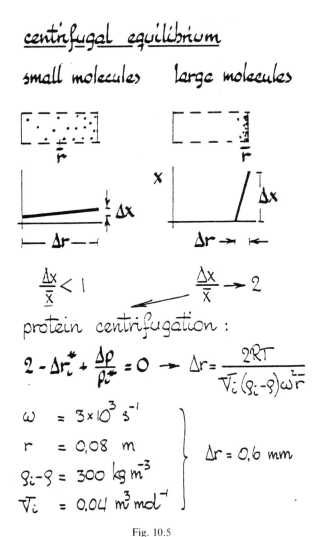

centrifugal equilibrium

small molecules   _large molecules_

$$\frac{\Delta x}{x} < 1 \qquad \frac{\Delta x}{x} \to 2$$

protein centrifugation :

$$2 - \Delta r_i^* + \frac{\Delta \rho}{\rho^*} = 0 \to \Delta r = \frac{2RT}{\overline{V_i}(\varrho_i - \varrho)\omega^2 r}$$

$$\omega = 3 \times 10^3 \ s^{-1}$$

$$r = 0.08 \ m$$

$$\varrho_i - \varrho = 300 \ kg \ m^{-3}$$

$$\overline{V_i} = 0.04 \ m^3 \ mol^{-1}$$

$$\Delta r = 0.6 \ mm$$

Fig. 10.5

SUPPORT FORCES

A third force which is often quite important is that due to the support
of solid matrices (Fig. 10.6). As an example we show the flow through
a filter cake. Here, the pressure gradient in the cake is supported by a
gradient of the stress in the cake. The cake in turn passes the force on
to a support grid. We make only one further remark on filtration: the
flow through the cake is almost solely viscous flow.

Support forces are important in pressure-driven membrane processes. You will not see them directly in the transport relations that
we use there. This is because they work on the solid matrix, and this is
just the component that we omit in the transport relations in membrane processes. The transport relations do, of course, contain the
resulting pressure gradient.

support force

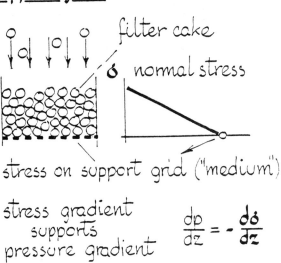

filter cake

normal stress

stress on support grid ("medium")

stress gradient supports pressure gradient

$$\frac{dp}{dz} = -\frac{d\sigma}{dz}$$

Fig. 10.6

ELECTRICAL FORCES

In an electrical field, charged species undergo electrical forces. These are proportional to the negative of the gradient of the electrical potential (Fig. 10.7). This potential is conventionally expressed in

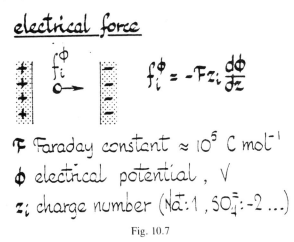

electrical force

$$f_i^\phi = -Fz_i\frac{d\phi}{dz}$$

F Faraday constant $\approx 10^5$ C mol^{-1}

ϕ electrical potential , V

z_i charge number (Na$^+$:1 , SO$_4^=$:-2 ...)

Fig. 10.7

volts. To obtain forces in molar units the gradient has to be multiplied by the Faraday constant. This constant has a large value; even small electrical potential differences can give rise to enormous forces. The force is also proportional to the charge number of the species. For simple electrolytes this number is small; much larger numbers are found with polyelectrolytes such as proteins and in charged micro-scopic particles.

DIFFERENCE APPROXIMATIONS

For transport through a 'film' also these forces can be brought in a dimensionless difference form similar to that used for the chemical potential gradient in Chapters 3 and 4. The results for the more important forces are shown in Fig. 10.8.

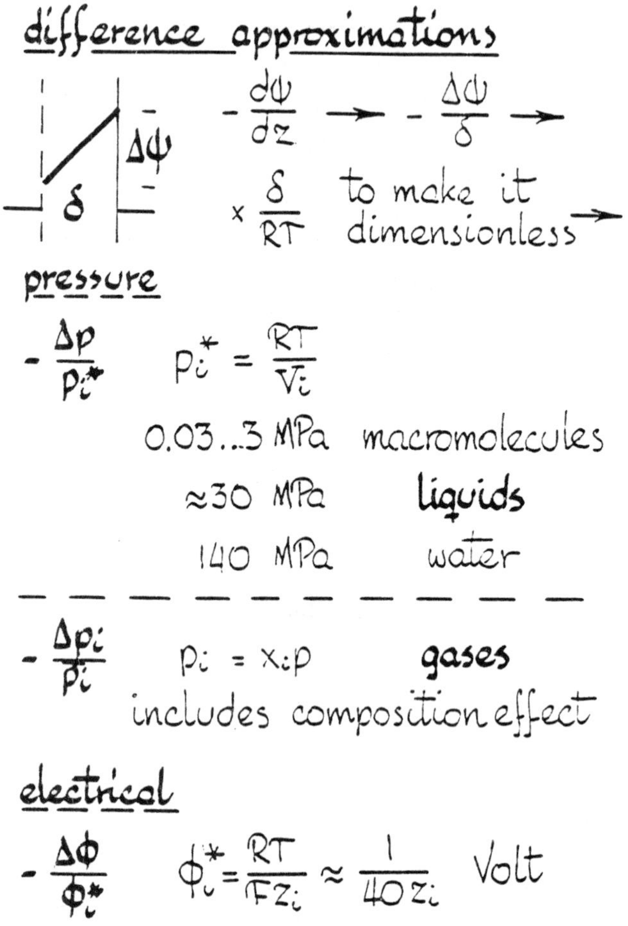

$$\underline{\textit{difference approximations}}$$

$$-\frac{d\psi}{dz} \longrightarrow -\frac{\Delta\psi}{\delta} \longrightarrow$$

$$\times \frac{\delta}{RT} \quad \text{to make it} \atop \text{dimensionless} \longrightarrow$$

$$\underline{\textit{pressure}}$$

$$-\frac{\Delta p}{p_i^*} \qquad p_i^* = \frac{RT}{V_i}$$

$$0.03...3 \text{ MPa} \quad \textbf{macromolecules}$$

$$\approx 30 \text{ MPa} \quad \textbf{liquids}$$

$$140 \text{ MPa} \quad \textbf{water}$$

$$-\frac{\Delta p_i}{p_i} \qquad p_i = x_i p \qquad \textbf{gases}$$

$$\text{includes composition effect}$$

$$\underline{\textit{electrical}}$$

$$-\frac{\Delta\phi}{\phi_i^*} \qquad \phi_i^* = \frac{RT}{Fz_i} \approx \frac{1}{40 z_i} \quad \text{Volt}$$

Fig. 10.8

The pressure force can be approximated by the pressure difference divided by a pressure constant. This constant contains the species volume and is different for each species. Macromolecules have a large volume, and pressure effects can be very significant. Water has an exceptionally small molar volume; very large pressure gradients are required to have any effect on it.

In gases the effects of composition and pressure gradients are best combined using partial pressures. The combined approximation is much more accurate than the sum of the two effects separately. Because of the large molar volume of gases, pressure forces are often important there.

The constant in the dimensionless electrical force has a value of only about one fortieth of a volt for a singly charged particle. So electrical forces can be very strong.

The total driving force on a species is simply the sum of the individual forces (Fig. 10.9).

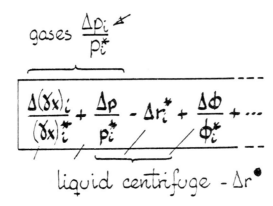

Fig. 10.9

SUMMARY

In this chapter we have seen a number of driving forces for mass transfer:

- gravity (a very weak force, except for large species),
- centrifugal forces,
- support forces, and
- electrical forces.

These 'external' forces usually give rise to pressure gradients. The pressure gradient is the sum of all external forces per volume of mixture. It is the cause of a pressure force, which is proportional to the species volume (partial molar volume).

For transport through a 'film' the forces can be brought into a dimensionless difference form (Fig. 10.9). You should memorize these 'forces'; we will be using them many times in the coming chapters.

11

Electrolytes

Now on to using the new forces, with a look at the diffusion of ions in electrolytes. Here, electrical fields play an important role. They can give rise to many unusual effects.

THE ELECTRONEUTRALITY RELATION

Electrolytes (Fig. 11.1) consist of a solvent (usually water) and dissolved positive and negative ions (cations and anions). Interfaces of electrolyte solutions nearly always carry a charge. This is due to several mechanisms: an important one is the preferred adsorption of certain ions at the interface. The surface charge attracts opposite charges, which form a more or less diffuse counter-layer. This 'double layer' is very thin — of the order of a few molecule diameters. The electrical field in the double layer is extremely high, and this gives rise to the most interesting phenomena in electrochemistry — surface reactions. These however, are not the subject of this book.

In mass transfer theory we are concerned with much thicker layers. A diffusion film has a thickness of the order of ten micrometres or, say, thirty thousand molecule diameters. The electrical potential gradient in this film is small (however, it is important for the transport of ions). As a consequence the charge separation in the diffusion film (and the bulk liquid) is negligible. For all practical purposes the concentrations of positive and negative charges are equal. This is known as the electroneutrality equation. This equation is essential in describing mass transfer in a way similar to that of the bootstrap relation.

TRANSPORT RELATIONS

The transport relations of ions in electrolytes are shown in Fig. 11.2. The most important difference with previous examples is the occur-

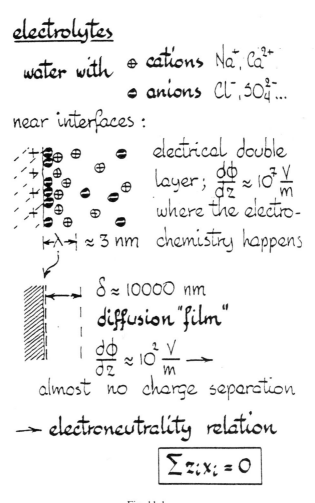

Fig. 11.1

rence of the electrical driving force. There are a few other things to be remarked.

- In electrochemistry, compositions are usually given in moles per litre. To stay consistent with the rest of this book we will, however, use mole fractions. For dilute solutions the mole fraction is one fifty-fifth of the concentration.

- In dilute solutions, activity coefficients of ions are given by the Debye–Hückel theory. This law is shown in Fig. 11.2 for single- and double-charged ions. The activity coefficients can be taken into account in the same way as in Chapter 8. As you can see, the changes of the coefficients with composition are not very large. In concentrated electrolyte solutions the variation of the activity coefficients is much larger, but you need more complicated thermodynamic models to describe this.

In very dilute solutions the activity coefficients become equal to

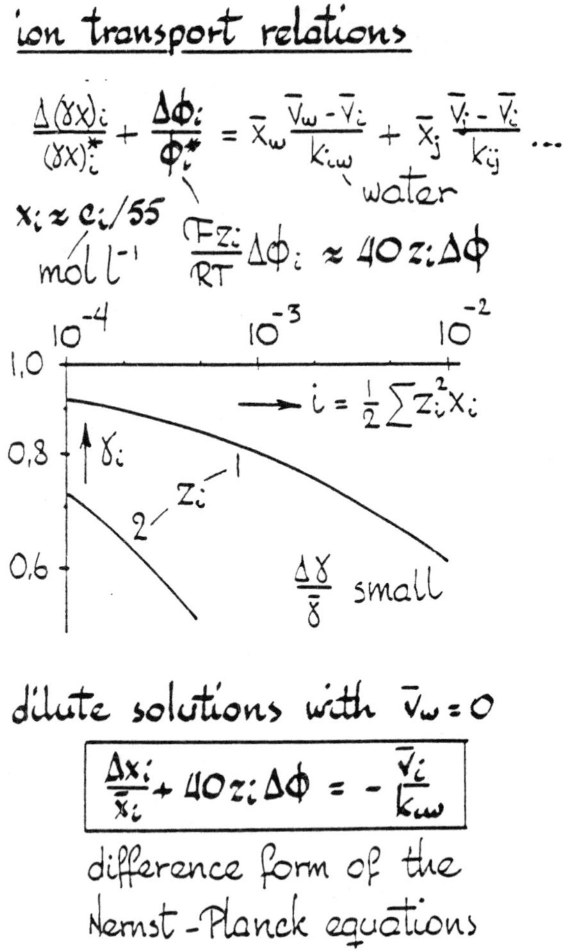

ion transport relations

$$\frac{\Delta(\delta x)_i}{(\delta x)_i^*} + \frac{\Delta\phi_i}{\phi_i^*} = \bar{x}_w \frac{\bar{V}_w - \bar{V}_i}{k_{iw}} + \bar{x}_j \frac{\bar{V}_j - \bar{V}_i}{k_{ij}} \ldots$$

water

$$x_i \approx c_i/55 \quad \frac{Fz_i}{RT}\Delta\phi_i \approx 40z_i\Delta\phi$$

mol l^{-1}

$$i = \tfrac{1}{2}\sum z_i^2 x_i$$

δ_i

z_i

$\frac{\Delta\delta}{\delta}$ small

dilute solutions with $\bar{V}_w = 0$

$$\boxed{\frac{\Delta x_i}{x_i} + 40z_i\Delta\phi = -\frac{\bar{V}_i}{k_{iw}}}$$

difference form of the
Nernst-Planck equations

Fig. 11.2

unity. In such solutions, friction between the ions and the solvent dominates. If the solvent velocity can also be taken as zero, the previous approximations lead to a difference form of what is known as the Nernst–Planck equation. We must stress here that the Nernst–Planck equation is just a special case of our general equations. It describes the simple examples we begin with quite adequately. In concentrated solutions you must go back to the general relations: this is not very difficult.

DIFFUSION OF HYDROCHLORIC ACID

The ions in this example (Fig. 11.3) have greatly differing diffusivities. Even so, hydrochloric acid diffuses in water as a single substance. How can this be?

The reason is not difficult to understand qualitatively. The hydrodiffusion of HCl in H₂O

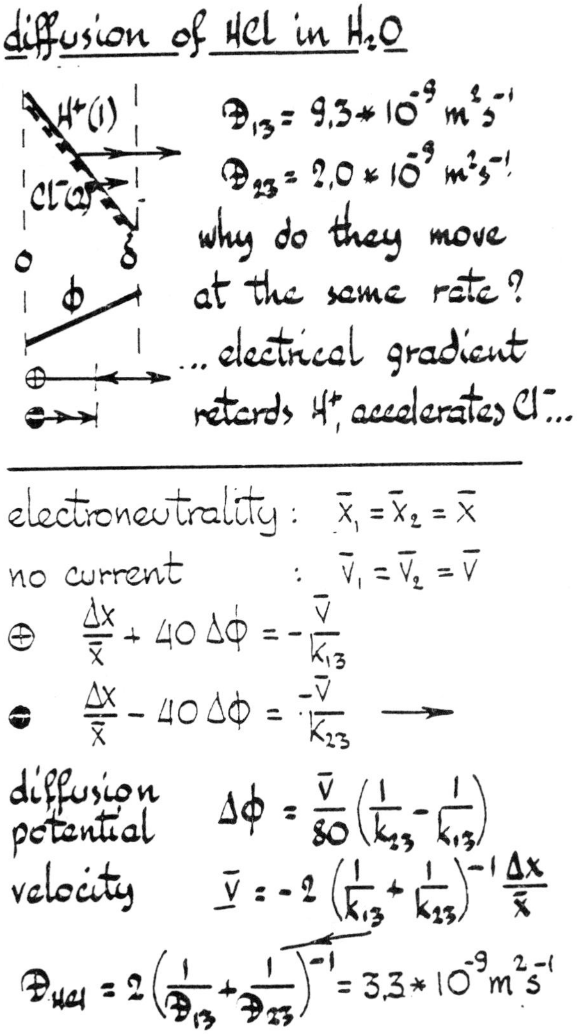

$$\mathcal{D}_{13} = 9.3 * 10^{-9}\ m^2 s^{-1}$$

$$\mathcal{D}_{23} = 2.0 * 10^{-9}\ m^2 s^{-1}$$

why do they move
at the same rate?
... electrical gradient
retards H⁺, accelerates Cl⁻...

---

electroneutrality : $\bar{x}_1 = \bar{x}_2 = \bar{x}$

no current   : $\bar{V}_1 = \bar{V}_2 = \bar{V}$

⊕ $\dfrac{\Delta x}{\bar{x}} + 40\ \Delta\phi = -\dfrac{\bar{V}}{k_{13}}$

⊖ $\dfrac{\Delta x}{\bar{x}} - 40\ \Delta\phi = \dfrac{-\bar{V}}{k_{23}}$ ⟶

diffusion
potential $\Delta\phi = \dfrac{\bar{V}}{80}\left(\dfrac{1}{k_{23}} - \dfrac{1}{k_{13}}\right)$

velocity $\bar{V} = -2\left(\dfrac{1}{k_{13}} + \dfrac{1}{k_{23}}\right)^{-1}\dfrac{\Delta x}{\bar{x}}$

$$\mathcal{D}_{HCl} = 2\left(\dfrac{1}{\mathcal{D}_{13}} + \dfrac{1}{\mathcal{D}_{23}}\right)^{-1} = 3.3 * 10^{-9}\ m^2 s^{-1}$$

Fig. 11.3

gen ion has a tendency to move more rapidly. This causes a minute charge imbalance (immeasurably small) and a potential gradient. This gradient slows down the hydrogen and speeds up the chloride so they go along together.

To describe this system we use the transport relations of the two ions and two other relations. They are the electroneutrality and 'no-current' relations. Solving the equations yields the diffusion potential difference and the species velocities. The diffusion potential is very small — in the order of millivolts. It is, however, essential in the transport relations. The transport of hydrogen chloride can be described by a single mass transfer or diffusion coefficient. Its value is intermediate those of the two ions separately.

... PLUS A TRACE OF SODIUM CHLORIDE

A salt, base or acid on its own in water diffuses as a single substance. You may be tempted to assume that mixed electrolytes also diffuse as separate substances. This can, however, lead to completely erroneous conclusions.

In Fig. 11.4 a trace of sodium chloride has been added to the

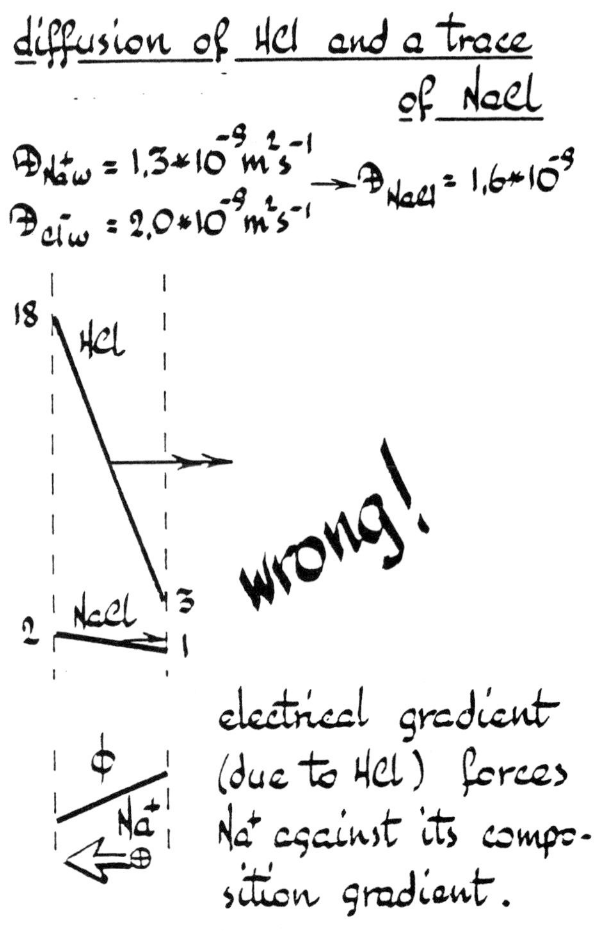

Fig. 11.4

previous example. The diffusivity of NaCl according to the previous formulae is given at the top of the figure. You might expect NaCl to diffuse down its gradient with a velocity corresponding to this diffusivity. This idea is wrong. In reality sodium will diffuse in the opposite direction! The movement of the sodium ion is determined by the electrical gradient caused by the hydrogen ion.

This is correctly predicted by the transport relations. As an exercise you might like to try this using a film thickness of ten micrometres.

POLARIZATION IN ELECTROLYSIS

In electrolysis an electrical field is used to separate ions from a solution (Fig. 11.5). There is a big difference compared with the

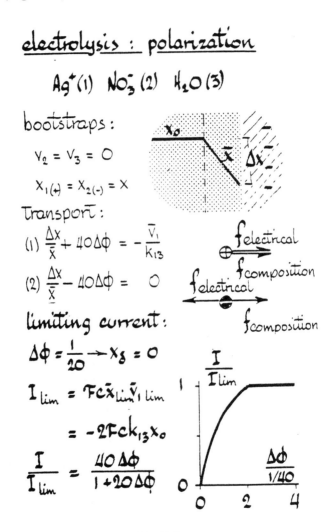

$$\underline{electrolysis : polarization}$$

$$Ag^+(1) \quad NO_3^-(2) \quad H_2O(3)$$

bootstraps:

$$V_2 = V_3 = 0$$

$$X_{1(+)} = X_{2(-)} = X$$

Transport:

$$(1) \quad \frac{\Delta X}{\bar{X}} + 40\Delta\phi = -\frac{\bar{V_1}}{k_{13}}$$

$$(2) \quad \frac{\Delta X}{\bar{X}} - 40\Delta\phi = 0$$

limiting current:

$$\Delta\phi = \frac{1}{20} \rightarrow X_\delta = 0$$

$$I_{lim} = F\bar{c}\bar{X}_{lim}\bar{V}_{1\,lim}$$

$$= -2Fck_{13}X_0$$

$$\frac{I}{I_{lim}} = \frac{40\Delta\phi}{1+20\Delta\phi}$$

Fig. 11.5

previous examples. There we applied a concentration difference and obtained an electrical potential difference. Here we apply an electrical field and obtain a concentration difference.

The electrolyte is a dilute solution of silver nitrate. We only consider the negative electrode. This is a fairly simple system: silver is deposited on the electrode. There are no other reactions. Nitrate does not partake in the reaction at this electrode. The bootstrap relations should be clear. The electroneutrality relation leads to the conclusion that the silver and nitrate concentrations must be the same everywhere. The transport relations for the two ions should be obvious.

From the transport relation of the nitrate it is seen that the ionic concentration at the electrode becomes zero when the potential difference is one twentieth of a volt. At this condition the silver transport rate reaches a maximum value. The corresponding charge transfer rate is known as the limiting current. A further increase of the potential difference does not lead to an increased current until another reaction — the electrolysis of water — starts taking over.

Note that the electrical and composition forces on the silver are both in the same direction. (For small forces you will find that this yields a flux equal to twice the product of the mass transfer coefficient and the concentration gradient.) The two forces on the nitrate ion just balance.

The effect of adding a considerable amount of an inert 'support' electrolyte (Fig. 11.6) is to almost eliminate the electrical potential

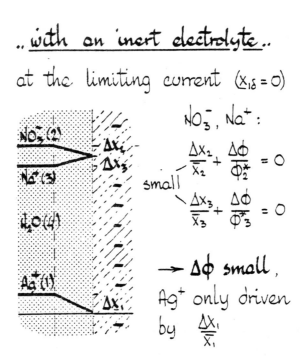

.. _with an inert electrolyte_ ..

at the limiting current $(x_{1s} = 0)$

$NO_3^-, Na^+:$

$$\frac{\Delta x_2}{\overline{x_2}} + \frac{\Delta \phi}{\overline{\phi_2^*}} = 0$$

small

$$\frac{\Delta x_3}{\overline{x_3}} + \frac{\Delta \phi}{\overline{\phi_3^*}} = 0$$

$\rightarrow \Delta \phi$ small,

Ag^+ only driven

by $\frac{\Delta x_1}{\overline{x_1}}$

Fig. 11.6

drop over the electrode film. It also halves the limiting current (the contribution of the electrical driving force disappears). Qualitatively the reasoning should be clear; details of the example can easily be followed numerically.

CONDUCTION AND FRICTION BETWEEN IONS

Fig, 11.7 shows an electrical conductivity cell. A hydrochloric acid

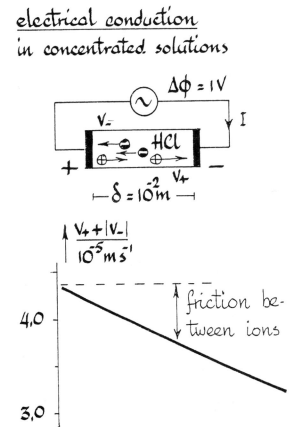

Fig. 11.7

solution is placed between two flat electrodes, one centimetre apart. The cell has the same cross-section throughout. The cell is fed by an alternating voltage with a root-mean-square value of one volt. If the frequency is sufficiently high you may assume that no concentration gradients develop in the cell.

From the current through the cell and the electrolyte concentration you can calculate the sum of the absolute values of the ionic velocities. If the Nernst–Planck relation were to apply, these should be independent of concentration. In reality a measurable decrease is noted with increasing concentration. This is due to friction between the ionic species, which are moving in an opposite direction. This is already noticeable at ionic mole fractions of well below 1%. (What is plotted is the square root of the mole fraction: this gives an almost straight line for reasons which will be explained. The mole fraction of ions in the middle of the figure is about 1%; it causes a reduction in the velocities of about 20%.) Clearly, the friction between ions is large compared to that between ions and water.

DIFFUSIVITIES IN ELECTROLYTE SOLUTIONS

There is a reasonable amount of information on ion–water diffusivities, and on the interactions between oppositely charged ions.

Ion–water diffusivities (Fig. 11.8) are fairly independent of con-

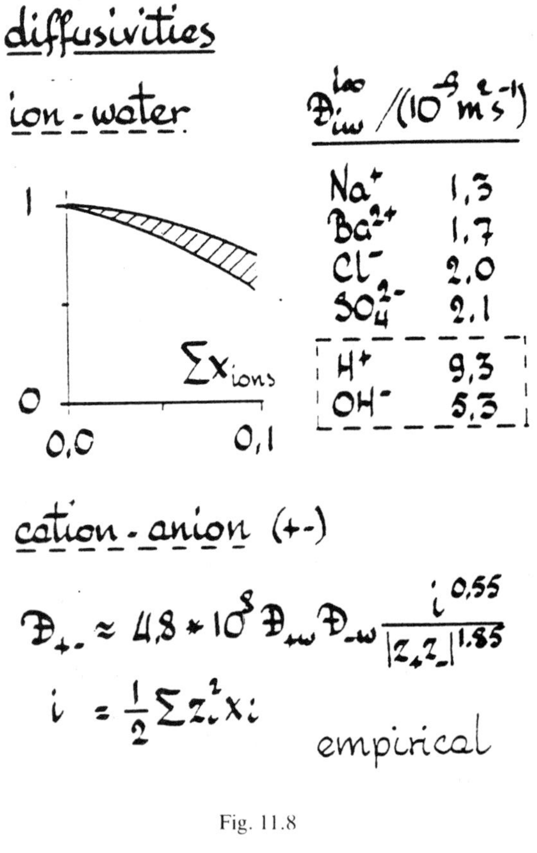

Fig. 11.8

centration. The values drop slowly at mole fractions in excess of a few per cent, mainly because of the increasing viscosity of the solutions. Ion diffusivities in dilute solutions can be found in several literature sources: a few values are given in the figure. The diffusivities of hydrogen and hydroxyl ions are considerably higher than those of other species.

A series of ion–ion diffusivities is plotted in Fig. 11.9. The

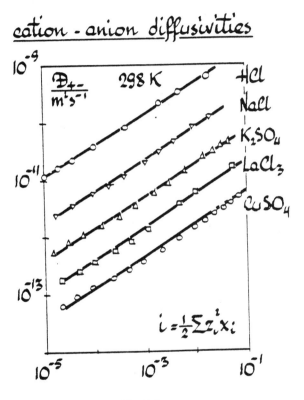

Fig. 11.9

parameter i along the horizontal axis is the mole fraction analogue of the 'ionic strength' used in electrolyte theory. For all components the diffusivity increases approximately with the square root of the ionic strength. (This, by the way, is the explanation of why the conductivity curve in Fig. 11.7 is a straight line.) The diffusivities are seen to decrease strongly with increasing charge numbers. Interactions between pairs of ions with the same charge numbers are almost identical.† All this indicates that ion–ion interactions are mainly due to electrostatic forces, as might be expected.

† With the exception again of pairs containing hydroxyl or hydrogen ions.

An empirical relation for ion–ion diffusivities is given at the bottom of Fig. 11.8. It is not very accurate. We have encountered deviations of up to 30%. Even then it can be useful, as ion–ion friction is usually not a too-large part of the total friction because of low ionic concentrations.

In the last part of this chapter we have introduced rather a large amount of material. If you would like some exercise on using it you might try to reconstruct Fig. 11.7. Remember that you do not need activity coefficients, as no composition gradients are involved in conduction.

SUMMARY

- The electrical driving force in electrolyte solutions gives rise to many new effects.
- It requires the use of a new relation: that for electroneutrality.
- Sometimes, multiple bootstrap relations are required. A new type is the 'no current' relation.
- The electrical field causes a strong coupling of the transport of ionic species. It causes the two constituents of a single salt to move as one component. In mixed salt solutions it may cause some species to move against their composition gradients.
- The relatively strong fields applied in electrolysis can cause depletion and limiting current phenomena.
- The friction between ions is quite high, and already noticeable at relatively low concentrations.

12

Membrane processes

This chapter gives a classification and some general remarks on membrane processes. We shall need these when discussing specific membrane applications.

A CLASSIFICATION

Membrane processes can be roughly classified according to the main driving force in the process (Fig. 12.1). The top row is mainly driven by composition differences, the second by an electrical field and the third by pressure gradients.

Of the composition-driven processes, dialysis is the oldest and most important. In dialysis the solvent (usually water) is the same on both sides of the membrane. Dialysis has found considerable application in the laboratory, in medical techniques and in the food industry. In pertraction, two different solvents are used. Pertraction has much in common with the process of liquid–liquid extraction. It is not (yet) of great significance.

Pervaporation is similar to evaporation. The process is operated with a pressure difference over the membrane. This difference is, however, much smaller than that in the pressure-driven processes which will be discussed later. At the pressure side is a liquid, at the other side a vapour. The membrane can greatly modify the selectivity of the evaporation process. Pervaporation is used for breaking azeotropes which cannot be separated by conventional distillation.

Of the electrically driven processes we will only discuss electrodialysis. Electrodialysis uses two kinds of membranes:

- positively charged membrane which only transfers negative ions and
- a negatively charged membrane which passes positive ions.

Electrodialysis is used on a fairly large scale, both for purifying and for concentrating electrolyte solutions.

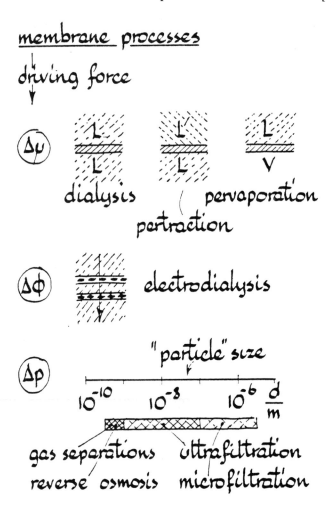

Fig. 12.1

Nowadays the most important membrane processes are those driven by pressure differences. These can be subdivided according to the size of the species that they separate. Small molecules (sizes smaller than about one nanometre) can be separated with tight (but very thin) polymer membranes. The differences between gas separations and liquid separations (known as reverse osmosis or as hyperfiltration) are sufficient to let them be regarded as different processes.

Molecules in the colloidal range (roughly between one and one hundred nanometres) are separated by ultrafiltration.

Ultrafiltration membranes are sieves, having fairly well defined pores. Ultrafiltration has found many applications in biotechnology and water treatment. Microfiltration retains again larger particles.

The distinction between ultrafiltration, microfiltration and conventional filtration is a little arbitrary. They all operate via the sieving action of:

- the membrane itself or
- a cake or gel layer formed on the membrane.

MEMBRANE STRUCTURES

There are many different types of membrane. For the purpose of this book the classification given in Fig. 12.2 is adequate.

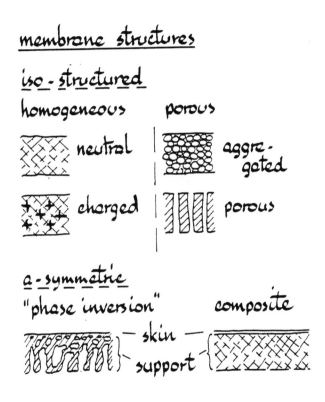

Fig. 12.2

The first group is formed by membranes which have the same structure throughout. The simplest are homogeneous films of a polymer. The polymer is usually chosen such that it swells in the solvent. The membrane thickness is of the order of a tenth of a millimetre.

The membranes used in electrodialysis are similar. The difference is that charged groups (either positive or negative) have been fixed to the polymer backbone. The counter-ions are mobile and can exchange with the surrounding solution.

Porous materials consisting of small sintered particles find many applications, and not only in membrane processes. Some porous materials have well-defined, more or less cylindrical pores.

The second group are the asymmetric membranes with a very thin skin and a relatively thick sponge-like support layer. These are the membranes used in pressure-driven processes. They are the ones which have brought membrane processes to the forefront. The actual membrane is the skin; the application determines the skin thickness and pore size.

Asymmetric membranes are made in two ways. In the first, a mixture of a solvent and the polymer is brought on to a support surface. This is then immersed in a bath of a liquid which does not dissolve the polymer. The solvent is extracted into the bath as the non-solvent diffuses into the polymer mixture. This results in demixing, with the polymer having a sponge structure. The demixing is rapid at the interface, and a skin is formed there with a fine structure. The second type is the composite membrane. It is made by forming a thin layer of a second polymer (or other material) on an open type of support.

SCALES IN MEMBRANE PROCESSES

In membrane processes we deal with phenomena happening on very different scales (Fig. 12.3). The membrane thickness is the largest scale at around a hundred micrometres in dialysis and electrodialysis. In ultrafiltration membranes the skin is around one micrometre, going down to one tenth of a micrometre in gas separations and reverse osmosis. The molecules or particles being separated are usually at least a factor of a thousand smaller than the membrane thickness.

MODELLING OF MEMBRANES

There are two fundamentally different ways of modelling permeable solid materials (Fig. 12.4). You should be aware which type of model you are using. Do not get the two mixed up!

- In the first the membrane is regarded as a heterogeneous structure. A simple example is the model of parallel cylindrical pores through a plate of an impermeable solid. This model is related to the models used for porous catalysts and adsorbents.
- In the second ('homogeneous') model we regard the membrane as a single phase. The components — including the solid matrix — are thought to form a molecular mixture.

Heterogeneous models are the natural choice for porous membranes. They allow some understanding of the influence of structure on membrane properties. Homogeneous models resemble

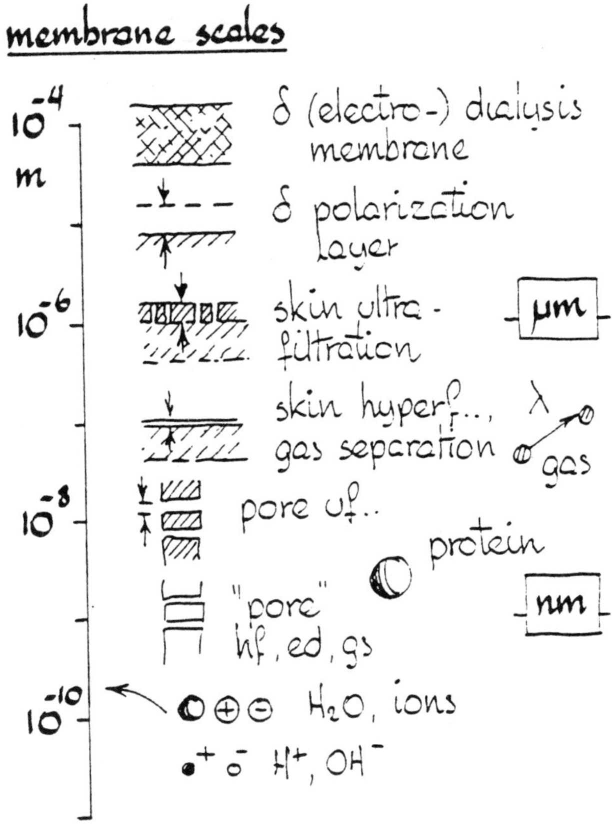

membrane scales

Fig. 12.3

closed polymer and gel membranes. However, for a given membrane, both models yield the same final results.

The species velocities in the two models are, of course, the same. In the pore model the concentration of a permeating species in a pore is higher than the concentration averaged along the membrane. However, transport only occurs in the pores, and these two effects cancel (as you can easily check).

EQUILIBRIA

At the membrane interface the membrane and the outer phase are assumed to be in equilibrium (Fig. 12.5). The ratio of the concentration of a species inside and outside the membrane at equilibrium is known as the volumetric distribution coefficient. If no preferential adsorption occurs it is equal to the void fraction or volume fraction of the permeants in the membrane.

The distribution coefficient is often fairly independent of composition. This is not so for charged membranes. Here the ion concentrations in the membrane are almost independent of the external

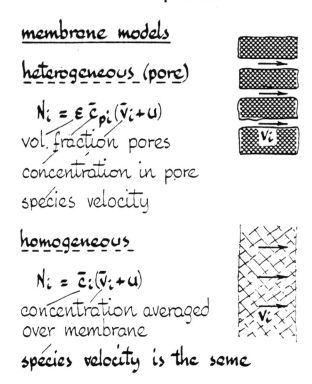

membrane models

heterogeneous (pore)

$$N_i = \varepsilon \bar{c}_{pi}(\bar{v}_i + u)$$

vol. fraction pores

concentration in pore

species velocity

homogeneous

$$N_i = \bar{c}_i(\bar{v}_i + u)$$

concentration averaged over membrane

species velocity is the same

Fig. 12.4

concentration. Changes of the external concentration are balanced by a change in the electrical potential difference between membrane and solution.

For the difference approximations we will be using, only the total potential difference over the membrane is required. Because we assume equilibrium at the membrane interfaces this can be calculated in two ways:

- from the changes inside the membrane and
- from the differences in potential just outside the two membrane surfaces.

The second method is often easier.

FRICTION

There is no such shortcut for the friction terms. Friction occurs inside the membrane and has to be calculated using mole fractions, velocities and diffusivities inside the membrane.

The membrane itself is one of the species to be considered in mass transfer calculations. You can define its mole fraction in different ways (Fig. 12.6).

- You may regard the membrane as a single molecule with mole fraction zero. (We have already seen how to do this in Chapter 4.)

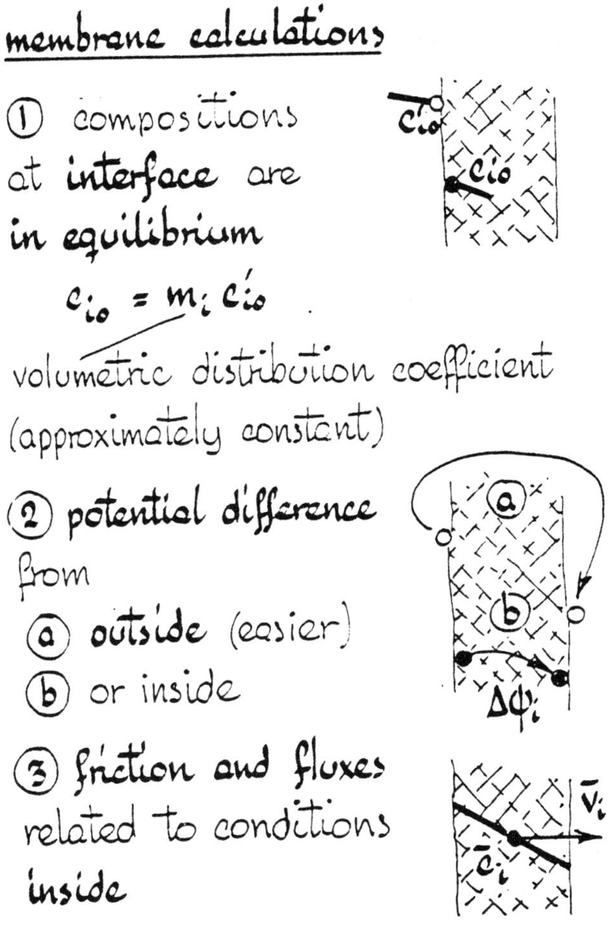

membrane calculations

① compositions
at **interface** are
in equilibrium

$$c_{io} = m_i c'_{io}$$

volumetric distribution coefficient
(approximately constant)

② potential difference
from
ⓐ outside (easier)
ⓑ or inside

③ friction and fluxes
related to conditions
inside

Fig. 12.5

This is done when using the heterogeneous model, but also in other cases.

- You may regard each monomer unit (or some fixed number of units) as a molecule.
- In charged membranes it is common to denote the membrane concentration by the concentration of fixed charges.

Some choices may be handier than others, however, they are all allowable. The choice you make does influence the numerical value of the diffusivities and mass transfer coefficients, so you must specify your choice.

SUMMARY

- In this chapter we have set up a classification of membrane processes according to the driving force: composition, electrical and pressure gradients.

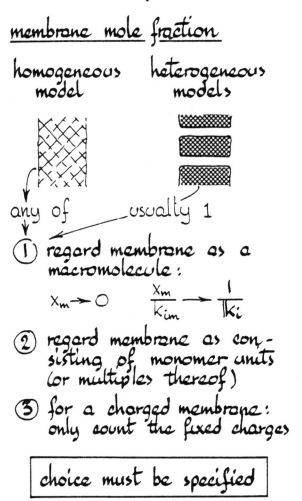

membrane mole fraction

homogeneous   heterogeneous
model   models

any of _____ usually 1

① regard membrane as a
macromolecule :

$$x_m \to 0 \qquad \frac{x_m}{K_{im}} \to \frac{1}{\overline{K_i}}$$

② regard membrane as con-
sisting of monomer units
(or multiples thereof)

③ for a charged membrane:
only count the fixed charges

| choice must be specified |

Fig. 12.6

- The composition-driven processes are further subdivided according to the phases on the two sides of the membrane. Dialysis has
the same solvent on both sides, pertraction two different liquids,
and pervaporation a liquid and a gas.
- The pressure-driven processes have been subdivided according to
the molecule or particle size separated. Small molecules (gases or
liquids) require gas separation or reverse osmosis. Macromolecules are separated by ultrafiltration and microscopic particles by
microfiltration.
- We have seen three ways of classifying membrane structures:

— as isostructured or as asymmetric,
— as homogeneous or heterogeneous and
— as electrically charged or neutral.

The most important are the non-charged asymmetric membranes.

The actual membrane there is a thin top layer or skin. The skin thickness is usually at least a thousand times the diameter of the species to be separated. The structure depends on the application.

- Two common models of membranes are the heterogeneous and the homogeneous ones. Both models yield the same final results. You should not get these models mixed up in intermediate calculations. Heterogeneous models can help you understand the effect of the structure of a membrane on its properties.
- Potential differences in membranes can be calculated either from the change in the outer or from the change in the inner conditions.
- Friction and flux calculations are related to concentrations, diffusivities and velocities inside the membrane. The values of the diffusivities depend on the way in which the membrane mole fraction is defined.

13

Gas permeation

The movement of gases through porous structures is reasonably well understood. It is also of considerable practical interest. This makes the subject a nice introduction to heterogeneous membrane models.

SEPARATION OF ISOTOPES

An important process in the nuclear industry is the separation of two isotopic forms of uranium hexafluoride (Fig. 13.1). These gases differ very little in properties, and their separation is difficult. One way is to make use of the small difference in friction that the two encounter passing through a porous membrane.

The membrane is assumed to consist of cylindrical pores, perpendicular to the membrane surface. Gas molecules collide with and rebound from the pore walls, but otherwise there are no interactions. (In particular, there is no adsorption.) As noted in Chapter 10 the driving forces in gas permeation contain differences of partial pressures. These are a combination of the effects of pressure and composition. However, in this example the change in composition over the membrane is very small and total pressures can be used instead.

Each transport equation contains two friction terms: one covering the interaction between the two gases, and the other friction between the gas and the solid. If either of the two is omitted we obtain one of the two classical diffusion equations:

- if the membrane friction is left out the equation for free binary diffusion,
- if the free diffusion term is left out the result is the Knudsen equation. This describes diffusion in solids where the pores are smaller than the mean free path of the gas.

The Maxwell–Stefan equations used here describe both phenomena

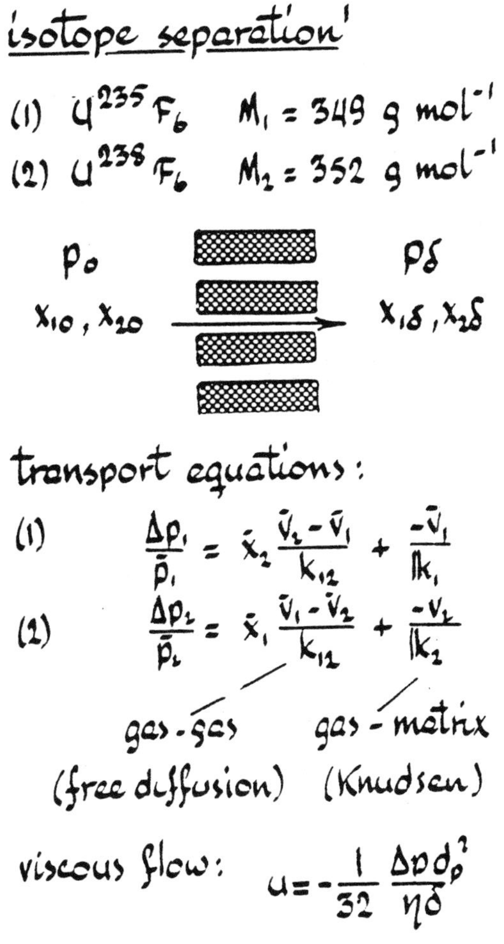

isotope separation[1]

(1) $U^{235} F_6$ $M_1 = 349$ g mol^{-1}

(2) $U^{238} F_6$ $M_2 = 352$ g mol^{-1}

p_0 .... $p\delta$

x_{10}, x_{20} ——————→ $x_1\delta, x_2\delta$

transport equations :

(1) $\dfrac{\Delta p_1}{\bar{p}_1} = \tilde{x}_2 \dfrac{\bar{v}_2 - \bar{v}_1}{k_{12}} + \dfrac{-\bar{v}_1}{\not{k}_1}$

(2) $\dfrac{\Delta p_2}{\bar{p}_2} = \tilde{x}_1 \dfrac{\bar{v}_1 - \bar{v}_2}{k_{12}} + \dfrac{-\bar{v}_2}{\not{k}_2}$

gas-gas   gas-matrix

(free diffusion) (Knudsen)

viscous flow: $u = -\dfrac{1}{32} \dfrac{\Delta p d_p^2}{\eta \delta}$

Fig. 13.1

and the transition from one mechanism to the other. Also important in this example is viscous flow. It can be predicted from the well-known equation of Poiseuille.

THE DIFFUSION COEFFICIENTS

We assume for simplicity that

- there is no adsorption at the pore walls and
- the pores have a diameter considerably larger than the gas molecules.

The gas–gas diffusivity is then equal to the value in free space† (Fig. 13.2).

The gas–wall diffusivity can be derived from momentum transfer

† In the heterogeneous models used here, the blocking by the solid material is taken into account via the void fraction of pores in the flux calculation.

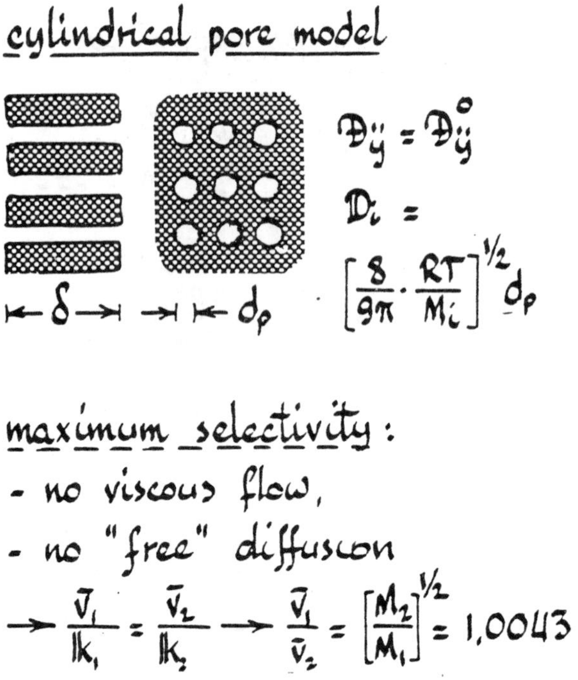

cylindrical pore model

$$\mathcal{D}_{ij} = \mathcal{D}_{ij}^{\circ}$$

$$D_i = \left[\frac{8}{9\pi} \cdot \frac{RT}{M_i}\right]^{1/2} d_p$$

maximum selectivity :

- no viscous flow,
- no "free" diffusion

$$\to \frac{\bar{v}_1}{k_1} = \frac{\bar{v}_2}{k_2} \longrightarrow \frac{\bar{v}_1}{\bar{v}_2} = \left[\frac{M_2}{M_1}\right]^{1/2} = 1.0043$$

Fig. 13.2

arguments; here we only show the final result. Gas–wall friction is seen to become more important for smaller pores and heavier gases.

As usual the mass transfer coefficients are equal to the ratio of the diffusivity and the 'film' (here membrane) thickness.

VELOCITY CALCULATIONS

Both free diffusion and viscous flow terms reduce the separation obtained by the membrane. So an upper limit for the separation can be obtained by neglecting these terms. The best separation is seen to be very small indeed. So large numbers of stages are required for any appreciable enrichment.

A sample calculation including all effects is given in Fig. 13.3. The separation is of course even less than the ideal value. If you wish you can play around with all the important parameters of the model:

• the pressure difference,
• the pore size,
• the membrane thickness and
• the temperature.

Indeed the model contains just about everything you would need on mass transfer to optimize a plant design. (Of course there are many other considerations involved in a complete design.)

We have omitted two mechanisms in this discussion:

isotope separation[2]

sample calculation

data	transport coeff's

$\bar{x}_1 = 0.50$

$\bar{x}_2 = 0.50$

$p_o = 500 \text{ kPa}$

$p_\delta = 400 \text{ kPa}$

$M_1 = 0.349 \text{ kg mol}^{-1}$

$M_2 = 0.352 \text{ kg mol}^{-1}$

$T = 300 \text{ K}$

$\mathcal{D}_{12}^o = 3.55 \times 10^{-7} \text{ m}^2\text{s}^{-1}$

$\eta = 30 \,\mu\text{Pa s}$

$\delta = 10^{-5} \text{ m}$

$d_p = 5 \times 10^{-9} \text{ m}$

transport coeff's

$k_{12} = 3.55 \times 10^{-2} \text{ ms}^{-1}$

$k_i = 2.2484 \times 10^{-2} \text{ ms}^{-1}$

$k_2 = 2.2388 \times 10^{-2} \text{ ms}^{-1}$

transport eq.'s

$$\frac{-100}{450} = 0.5 \frac{\bar{v}_2 - \bar{v}_1}{k_{12}} - \frac{\bar{v}_1}{k_1}$$

$$\frac{-100}{450} = 0.5 \frac{\bar{v}_1 - \bar{v}_2}{k_{12}} - \frac{\bar{v}_2}{k_2}$$

results

$\bar{v}_1 = \quad 4.9923 \text{ mm s}^{-1}$

$\bar{v}_2 = \quad 4.9792 \text{ mm s}^{-1}$

$u = \quad 0.2604 \text{ mm s}^{-1}$

$$\frac{(\bar{v}_1 + u)}{(\bar{v}_2 + u)} = 1.0025$$

Fig. 13.3

- Surface diffusion. This mechanism can be important with more condensable gases which adsorb in high concentrations on the pore walls.
- Configurational diffusion, which is important when the pore diameters are reduced towards the molecular diameters. This mechanism is very important in zeolites.

LOOKING BACK

Playing a bit with the cylindrical pore model will give you more understanding of two examples in Chapter 2.

A fairly fundamental way of describing the 'three gases' example is to consider it as a problem with four components:

- the three gases and

- the capillary wall.

If the volume of the bulbs is large compared with that of the capillary, you may regard the flow at any instant as stationary. You can then set up the transport equations for the three gases, including wall friction and a pressure gradient. This of course causes viscous flow. The geometry also indicates that the net flow must be close to zero. If you solve the set of equations you will find a minute pressure difference, because the hydrogen migrates more rapidly than the other gases. It is this difference which balances the flows.

A simpler way of describing the problem is to regard it as free diffusion of three gases with a 'no net flow' bootstrap relation. For not-too-narrow capillaries you will find that the results of the two approaches are almost identical.

The cylindrical pore model also gives a good description of the experiment with 'two gases and a porous plug'. If there is no pressure difference over the plug, the mole fraction differences of the two gases are equal but opposite. It is then easy to show that the ratio of the fluxes is determined solely by the gas–matrix friction terms. These in turn contain the square root of the molar mass of the two gases, as we had already seen.

A BED OF SPHERES

Many porous media are more realistically described as aggregates of spherical particles. Models for such systems have also been developed.

Fig. 13.4 shows a few aspects of the gas–gas diffusivity. The idea is that we are dealing with tortuous pores with wide parts and also constrictions. The tortuosity has two effects: it increases the pore length and reduces the effective pore width. The constrictions also reduce the effective diffusivity: the extra friction in the narrow parts is not balanced by the reduced friction in the wide parts. All in all the effect is to reduce the gas–gas diffusivity by roughly one order of magnitude (less in very open structures, more in compressed structures).

In many sources you will find the ratio of tortuosity squared over constrictivity designated as the 'tortuosity'. We have kept the two things separate. Note that the gas–gas coefficient does not depend on the size of the particles in the membrane.

The gas–matrix (Knudsen) diffusivity is covered in Fig. 13.5. Possibly to your surprise there is a simple theory which describes the behaviour of this coefficient quite well.

In the following you may think your leg is being pulled. But really it works. The idea is that the matrix particles can be regarded as enormous molecules ('dust') in an ideal gas mixture (!). Because they are so large their thermal motion is negligible and that is just what happens in a sintered matrix.

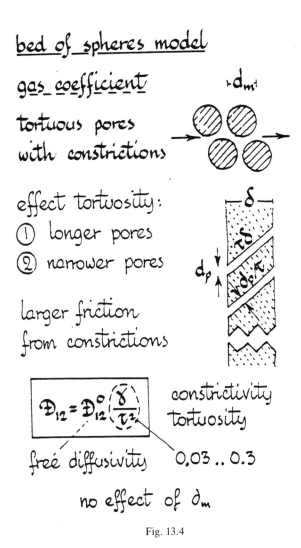

Fig. 13.4

For very large differences in molecule sizes the kinetic gas theory provides the binary gas diffusivity given in the first formula. With a little geometry it is easy to obtain the mole fraction of matrix particles. Dividing the binary diffusivity by this value gives the gas–membrane coefficient. This is seen to increase

- linearly with the coarseness (particle size) of the membrane,
- with the square root of the absolute temperature and
- inversely with the square root of the molar mass of the diffusing species.

All these observations are borne out in practice. The viscous flow in a matrix of spherical particles can be calculated from relations for packed beds (Fig. 13.6). The viscous flow increases with the pressure gradient and the square of the diameter of the matrix particles. The

matrix coefficient.

(dusty gas model)

matrix taken as
extremely large gas
(dust) molecules (!)

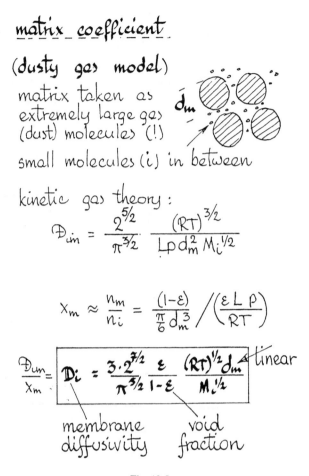

small molecules (i) in between

kinetic gas theory :

$$\mathcal{D}_{im} = \frac{2^{5/2}}{\pi^{3/2}} \cdot \frac{(RT)^{3/2}}{L\rho d_m^2 M_i^{1/2}}$$

$$X_m \approx \frac{n_m}{n_i} = \frac{(1-\varepsilon)}{\frac{\pi}{6}d_m^3} \Big/ \left(\frac{\varepsilon L \rho}{RT}\right)$$

$$\frac{\mathcal{D}_{im}}{X_m} = \boxed{D_i = \frac{3 \cdot 2^{3/2}}{\pi^{3/2}} \frac{\varepsilon}{1-\varepsilon} \frac{(RT)^{1/2} d_m}{M_i^{1/2}}} \text{ linear}$$

membrane   void
diffusivity   fraction

Fig. 13.5

velocity we need is not the superficial velocity, but the component of the pore velocity perpendicular to the membrane. That is why the exponent in the void fraction is one less than in packed bed relations.

THE DUSTY GAS MODEL

The results of the bed-of-spheres model are very similar to those of the model with cylindrical pores. The two models are examples of a more general model known as the 'dusty gas model' (Fig. 13.6). The constants in the transport coefficients are usually determined experimentally.

SUMMARY

This chapter has introduced two heterogeneous models for transport of gases through a porous medium:

- a model with parallel cylindrical pores and

viscous flow

squared

$$u = \frac{-1}{170} \cdot \frac{\varepsilon^2}{(1-\varepsilon)^2} \cdot \frac{\Delta p \, d_m^2}{\eta \delta}$$

average velocity in pores

dusty gas model

$$\mathcal{D}_{ij}'' = constant^1 \cdot \mathcal{D}_{ij}^0$$

$$D_i = constant^2 \cdot \left[\frac{T}{M_i}\right]^{1/2}$$

$$u = constant^3 \cdot \frac{[-\Delta p]}{\eta \delta}$$

determined by experiment
for each material

Fig. 13.6

- a model consisting of a packed bed of equal spheres.

Both models are similar. They contain terms for

- gas–gas interactions,
- gas–wall interactions and
- viscous flow.

They both require some information on the membrane structure:

- the void fraction and
- a pore diameter or particle size.

In addition the bed-of-spheres model requires the tortuosity and constrictivity of the membrane.

The two models can be regarded as special cases of a more general model. This is known as the 'dusty gas model'. The coefficients of this model are usually determined experimentally.

14

Composition forcing

Here we consider two processes: dialysis and pervaporation. Both use polymer membranes, which will be regarded as homogeneous.

DIALYSIS

In the example in Fig. 14.1, urea is being removed from a solution in water also containing a protein. The protein is thought to be inert and not to influence the transport process. Urea will be diffuse from left to right, and water from right to left.

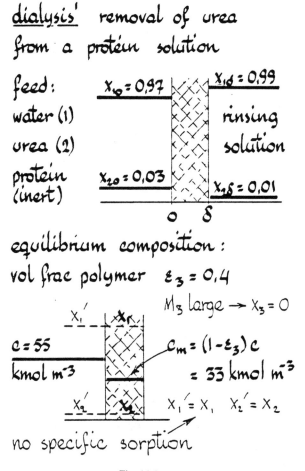

Fig. 14.1

Before doing the transport calculations we first look at the membrane equilibria. The membrane is swollen: it only contains a volume fraction polymer of 0.4. The molar mass of the membrane material is very high, and the membrane mole fraction is taken as zero. In such open membranes the assumption of equal distribution coefficients of the components is a reasonable one. With a volume fraction of the permeants of 0.6 this means that all concentrations inside the membrane are six tenths of those outside.

The transport coefficients in the membrane are related to diffusivities. Fig. 14.2 gives you an impression of how these might behave as a function of the volume fraction solvent in the membrane. (This can be influenced by the choice of polymer, the cross-linking and also the way in which the membrane is formed.) The lower the solvent volume fraction, the tighter the membrane becomes. In open membranes, friction between the solvent and solute is important: the solvent–solute diffusivity is lower than the membrane diffusivities. It does not decrease very rapidly in tighter membranes, however, and in that case friction with the membrane becomes much more important. In

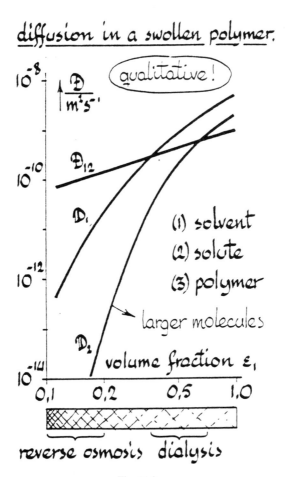

Fig. 14.2

tight membranes the membrane diffusivities decrease very strongly as the volume fraction solvent decreases. This is even more pronounced for large solute molecules. The picture given here is qualitative: we do not have complete data on any solvent–solute–polymer system!

Let us return to the transport equations for dialysis (Fig. 14.3). With the given membrane thickness and taking diffusivities from the previous figure the mass transfer coefficients are easily computed. They are of the order of a few micrometres per second. The interactions with the polymer are given by the membrane coefficients we introduced in Chapter 4. The transport equations are otherwise very similar to those in earlier examples and should present no difficulties.

$$\underline{dialysis}^2$$

transport coefficients

$$\delta = 10^{-4} m$$

$$\mathcal{D}_{12} = 5 \times 10^{-10}\, m^2 s^{-1} \longrightarrow k_{12} = 5\, \mu m\, s^{-1}$$

$$\mathcal{D}_1 = 10 \times 10^{-10} \longrightarrow k_1 = 10$$

$$\mathcal{D}_2 = 2 \times 10^{-10} \longrightarrow k_2 = 2$$

MS-eqs :

$$(1)\quad \frac{0.02}{0.98} = 0.02\, \frac{\bar{v}_2 - \bar{v}_1}{5} + \frac{v_m - \bar{v}_1}{10}$$

$$(2)\quad \frac{-0.02}{0.02} = 0.98\, \frac{\bar{v}_1 - \bar{v}_2}{5} + \frac{v_m - \bar{v}_2}{2}$$

bootstrap $v_m = 0$

$$\longrightarrow \bar{v}_1 = -0.14\, \mu m\, s^{-1} \qquad \bar{v}_2 = 1.4\, \mu m\, s^{-1}$$

$$N_1 = \bar{v}_1 \bar{x}_1 c_m = -0.14 \times 0.98 \times 33\; mmol\, m^{-2} s^{-1}$$

$$N_2 = \bar{v}_2 \bar{x}_2 c_m = +1.4 \times 0.02 \times 33 \dots$$

Fig. 14.3

PERVAPORATION

In biotechnology, fermentation rates are frequently controlled by product inhibition This example (Fig. 14.4) is from an attempt to avoid this by removal of the product (butanol) from a fermentation broth. The product was removed continuously by pervaporation using a silicone rubber membrane.

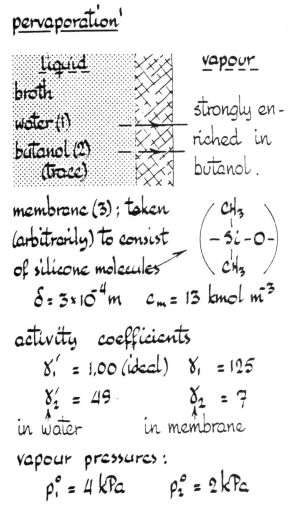

Fig. 14.4

Here we take the membrane to consist of a solution of water, butanol and the silicone groups forming the polymer. (How you describe the composition is arbitrary. Of course the choice does influence the numerical values of mole fractions, distribution coefficients, and diffusivities.) This yields a total membrane concentration of 13 kmol m^{-3}. The activity coefficients of water and butanol in the liquid and in the membrane have been determined experimentally. The vapour pressures are from a handbook.

   With the above data the equilibria at the membrane interfaces can
be constructed (Fig. 14.5). This is all standard thermodynamics. A
few points are to be noted. The total fraction of butanol and water in
the membrane is very low. This allows the butanol/water frictional
interaction to be neglected. In a relative sense butanol is much more
soluble than water. (Its distribution coefficient is about nine hundred
times larger). Clearly the butanol is more compatible than water with
the non-polar nature of the silicone rubber. Also the downstream
compositions are not given — only the total downstream pressure.
Because of this the transport relations have to be solved by trial and
error. We did this with an equation-solving package on a personal
computer.
   The diffusivities and mass transfer coefficients were in this case
determined experimentally. The value for water is quite high — that

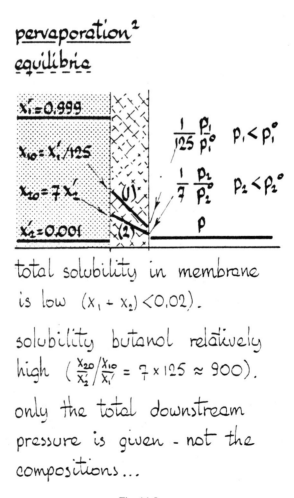

$$\text{pervaporation}^2$$
$$\text{equilibria}$$

$x_1' = 0.999$

$x_{10} = x_1'/125$ $\frac{1}{125}\frac{p_1}{p_1^o}$ $p_1 < p_1^o$

$x_{20} = 7 x_2'$ $\frac{1}{7}\frac{p_2}{p_2^o}$ $p_2 < p_2^o$

$x_2' = 0.001$ (2) p

total solubility in membrane
is low $(x_1 + x_2) < 0.02)$.

solubility butanol relatively
high $(\frac{x_{20}/x_{10}}{x_2'/x_1'} = 7 \times 125 \approx 900)$.

only the total downstream
pressure is given - not the
compositions ...

Fig. 14.5

for butanol about twenty times lower. The reason for the high water diffusivity is not the swelling of the polymer. The silicone rubber is close to its melting trajectory and the polymer chains are very mobile.

Together with the boundary conditions there are twelve equations to be solved, of which several are non-linear (Fig. 14.6).

pervaporation[3]

transport parameters

$$\mathcal{D}_{13} = 3 \times 10^{-10}\, m^2 s^{-1} \longrightarrow k_{13} = 1\ \mu m\, s^{-1}$$

$$\mathcal{D}_{23} = 1.8 \times 10^{-11} \qquad\qquad k_{23} = 0.06$$

all equations

equilibria $\quad x_{10} = x_1'/125 \qquad x_{20} = 7 x_2'$

$$x_1\delta = p_1/(125\, p_1^o) \quad x_2\delta = p_2/(7 p_2^o)$$

auxiliary $\quad \Delta x_1 = x_1\delta - x_{10} \qquad \Delta x_2 = x_2\delta - x_{20}$

$$\bar{x}_1 = (x_1\delta + x_{10})/2 \quad \bar{x}_2 = (x_2\delta + x_{20})/2$$

transport
+ bootstrap $\quad \dfrac{\Delta x_1}{\bar{x}_1} = x_3 \dfrac{-\bar{v}_1}{k_{13}} \quad \dfrac{\Delta x_2}{\bar{x}_2} = x_3 \dfrac{-\bar{v}_2}{k_{23}}$

extra
constraints $\quad p_1 + p_2 = p \quad \dfrac{p_1}{p_2} = \dfrac{\bar{v}_1\, \bar{x}_1}{\bar{v}_2\, \bar{x}_2}$

12 equations − 12 unknowns − non-
linear − solved numerically →
then $\qquad N_1 = \bar{v}_1\, c_m\, \bar{x}_1 \quad N_2 = \bar{v}_2\, c_m\, \bar{x}_2$

Fig. 14.6

A set of results with varying downstream pressures is given in Fig. 14.7. Permeation only becomes measurable at pressures lower than the vapour pressure of the feed. The water flux increases linearly as the pressure is decreased. The butanol flux increases more than linearly so a higher butanol selectivity is obtained at low downstream pressures.

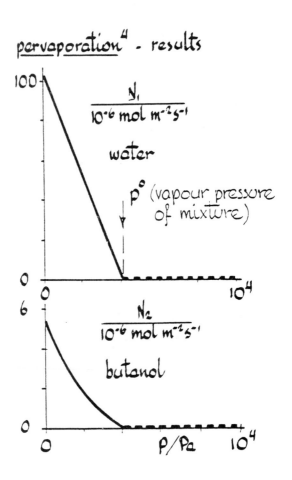

Fig. 14.7

A FEW FINAL REMARKS

- The driving force in pervaporation is the potential difference between the saturated liquid and the vapour. (The difference between the actual liquid and the saturated liquid is negligible).
- The selectivity towards butanol is not a kinetic effect: it is due to the difference in the solubilities of water and butanol in the membrane.
- In other pervaporation systems, concentrations in the membrane can be much higher than here. You must then expect diffusivities to vary over the membrane, as the swelling is higher on the upstream side.

There is no summary at the end of this and the following chapter. We will look at all membrane processes in the summary of Chapter 16.

15

Electrical forcing

ELECTRODIALYSIS

Electrodialysis is used for removing electrolytes from solution and for concentrating electrolytes. It is slightly more complicated than other membrane processes in that it uses two kinds of membrane. (Fig. 15.1). Both consist of a polymer backbone — often polystyrene — but nowadays also fluorinated hydrocarbons. One type has fixed negative charges such as sulphonic acid groups; the other has fixed positive charges such as quaternary ammonium groups. The negative membrane only transfers positive ions and vice versa.

The membranes are arranged in stacks of several hundred. These alternate: positive–negative–positive etc. Liquid flows in thin layers between the membranes. These layers are alternately connected to form two sets of compartments. In the figure, four membranes are shown, but do realize that this is certainly not to scale.

Across the membrane stack an electrical field is applied. This forces ions through the membranes, alternately diluting and concentrating the electrolyte between the membranes.

Before looking at transport in the membranes we first look at the compositions in and around the membrane (Fig. 15.2) and at the membrane equilibria (Fig. 15.3). We only consider the negative membrane — the principles are equal for the other positive one. Water has a molar concentration of 55 mol per litre. It contains only small amounts of the electrolyte (here, sodium chloride).

The membrane composition is characterized once again in a different manner. Here we choose the fixed charge concentration as the membrane concentration. Owing to the polar nature of the fixed charges, the membrane swells in water. The swelling is, however, limited by cross-linking of the polymer. Overall, about three quarters of the volume is occupied by membrane material. The chloride 'co-ion' is effectively excluded from the membrane by the fixed charges. (We shall look at this in a moment.) So the sodium concentration

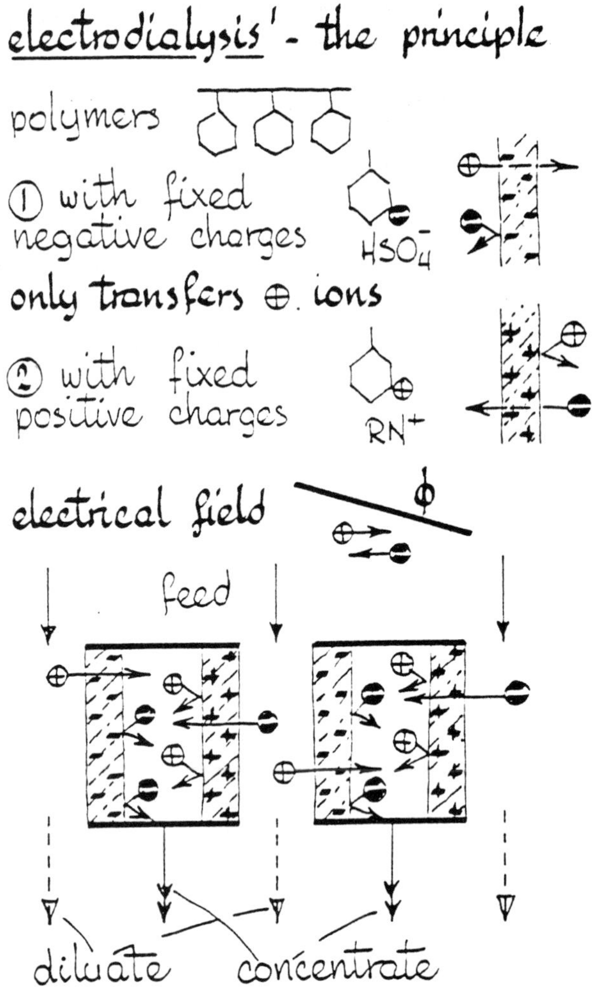

electrodialysis' - the principle

polymers

① with fixed negative charges HSO_4^-

only transfers ⊕ ions

② with fixed positive charges RN^+

electrical field ϕ

feed

diluate   concentrate

Fig. 15.1

must be equal to the fixed charge concentration. Together these yield a concentrated electrolyte solution in the membrane, quite different from the external solution.

Fig. 15.3 gives a brief summary of the ionic equilibria of the membrane. It is assumed here that the membrane is homogeneous and that the liquids are ideal. These assumptions do not hold accurately and the results are only approximate.

The sodium in the membrane tends to diffuse outward. This causes a potential difference of the order of a tenth of a volt. (You obtain the equilibrium relation at the top of Fig. 15.3 by setting the driving force in our transport relations equal to zero.)

The potential difference excludes the chloride ions in a similar manner. Combining the two relations shows that the chloride concentration in the membrane varies with the square of the external

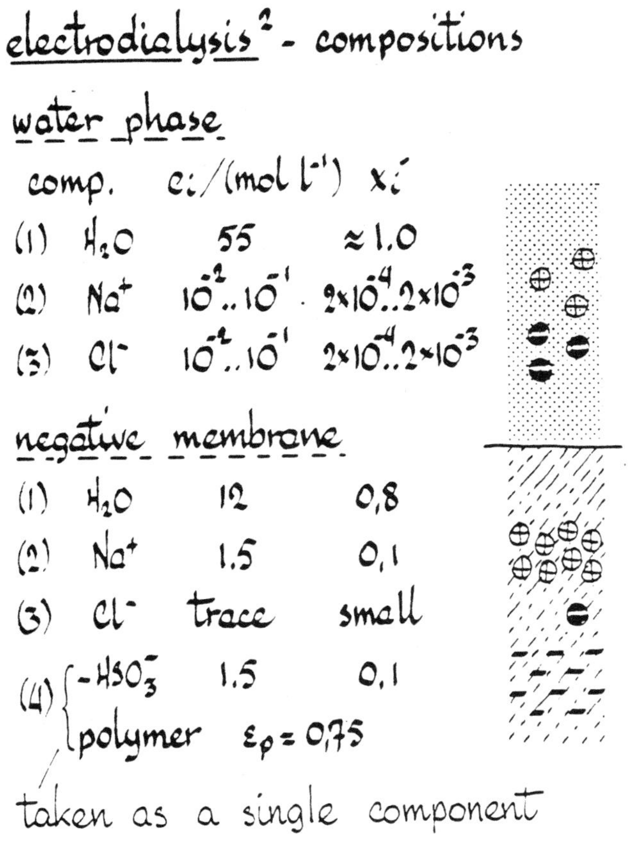

electrodialysis[2] - compositions

water phase

comp. c_i / (mol l^{-1}) x_i

(1) H_2O 55 ≈ 1.0

(2) Na^+ $10^{-2}..10^{-1}$. $9\times10^{-4}..2\times10^{-3}$

(3) Cl^- $10^{-2}..10^{-1}$ $2\times10^{-4}..2\times10^{-3}$

negative membrane

(1) H_2O 12   0,8

(2) Na^+ 1.5   0,1

(3) Cl^- trace   small

(4) $\begin{cases} -HSO_3^- & 1.5 & 0,1 \\ polymer & \varepsilon_\rho = 0,75 \end{cases}$

taken as a single component

Fig. 15.2

concentration. At low external concentration the chloride exclusion is very effective. There are three species in the membrane: water, sodium and the membrane material. The water–sodium and water-–membrane diffusivities are about one order of magnitude lower than in free solution (Fig. 15.4). This is mainly due to geometrical effects such as we have seen in the bed-of-spheres model. The sodium–membrane diffusivity is considerably lower: just as in free solutions the friction between ions is very high. Note the extremely low value of the sodium–membrane mass transfer coefficient.

A set of transport calculations is given in Fig. 15.5. The concentration to the right of the membrane has been chosen ten times higher than at the left. The applied potential difference is two tenths of a volt.

This is an example where the species potential difference is most easily calculated from the changes between points just outside the membrane. As you can see, the driving force on the water is almost zero. Water is propelled by friction with the moving sodium ions: its velocity is intermediate that of sodium and the membrane. The main

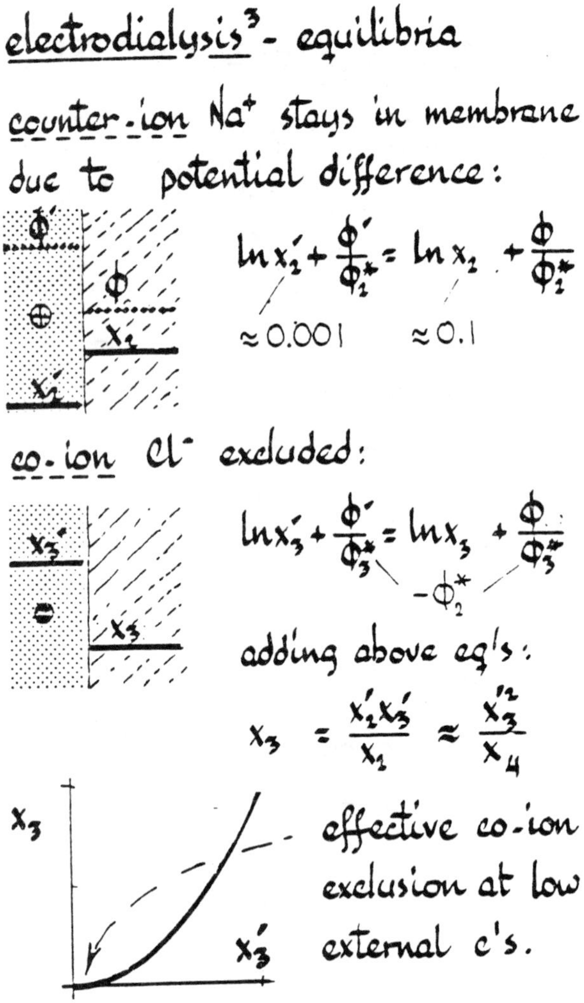
electrodialysis³ - equilibria
counter-ion Na+ stays in membrane
due to potential difference:
ln x'2 + φ'/φ*2 = ln x2 + φ/φ*2
≈ 0.001
≈ 0.1
co-ion Cl- excluded:
ln x'3 + φ'/φ*3 = ln x3 + φ/φ*3
-φ*2
adding above eq's:
x3 = x'2x'3/x2 ≈ x'3²/x4
x3
x'3
effective co-ion exclusion at low external c's.

Fig. 15.3

driving force on sodium is the electrical force. The composition driving force is, however, not negligible.

Before leaving the subject of electrodialysis, a few additional remarks are in order.

- Polarization is very important. The effects are similar to those in electrolysis.
- At high external electrolyte concentrations the chloride concentration in the negative membrane does start to have effect.

Both effects can well be taken into account by our model. This does not imply that building a complete model of an electrodialysis stack is a trivial matter: we know better from experience.

<u>eldial'sis[4]</u> - transport parameters

H_2O / Na^+ $\mathcal{D}_{12} = 1.4 \times 10^{-10} \, m^2 s^{-1}$

H_2O / HSO_3^- $\mathcal{D}_{14} = 2.0 \times 10^{-10} \, m^2 s^{-1}$

$\approx 10 \times$ lower than in free solution
(due to geometrical obstruction)

Na^+ / HSO_3^- $\mathcal{D}_{24} = 1.3 \times 10^{-11} \, m^2 s^{-1}$

large friction between ions ;
$\mathcal{D} \approx 20 \times$ lower than $\mathcal{D}$ free.

$\delta = 2 \times 10^{-4} m$ $k_{ij} = \mathcal{D}_{ij} / \delta$

$\longrightarrow k_{12} = 0.7 \times 10^{-6} \, m s^{-1}$

$k_{14} = 1.0 \times 10^{-6} \, m s^{-1}$

$k_{24} = 6.5 \times 10^{-8} \, m s^{-1}$

Fig. 15.4

electrodialysis[5] - transport

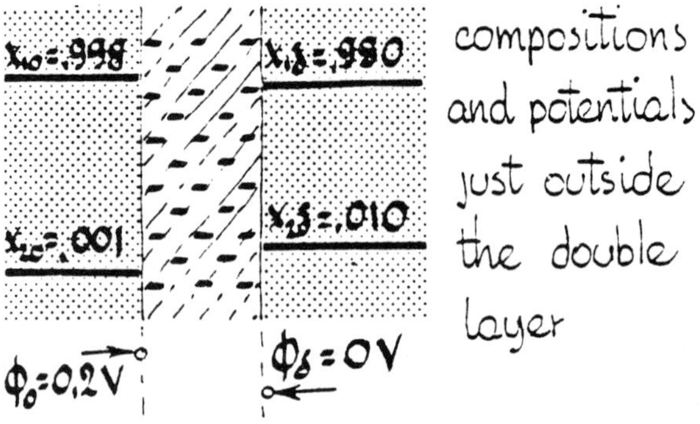

$x_{10} = .998$ $x_{1\delta} = .990$ compositions and potentials just outside the double layer

$x_{20} = .001$ $x_{2\delta} = .010$

$\phi_0 = 0.2\,V$ $\phi_\delta = 0\,V$

H_2O: $\dfrac{\Delta x_1'}{\bar{x}_1'} = \bar{x}_2\dfrac{\bar{v}_2-\bar{v}_1}{k_{12}} + \bar{x}_4\dfrac{-\bar{v}_1}{k_{14}}$

≈ 0

Na^+: $\dfrac{\Delta x_2'}{\bar{x}_2'} + \dfrac{\Delta\phi}{1/40} = \bar{x}_1\dfrac{\bar{v}_1-\bar{v}_2}{k_{12}} + \bar{x}_4\dfrac{-\bar{v}_2}{k_{24}}$

≈ 2 ≈ -8

$\bar{v}_1 = 1.96$ $\bar{v}_2 = 3.21 \ \mu m \ s^{-1}$

$N_1 = 23.6$ $N_2 = 4.82 \ mmol \ m^{-2} s^{-1}$

Fig. 15.5

16

Pressure forcing

In this chapter we have a look at two membrane processes in which pressure gradients play an important role:

- desalination of water by reverse osmosis and
- concentration of proteins by ultrafiltration.

The chapter ends with an overview of all membrane processes considered in this book.

REVERSE OSMOSIS

The production of potable water from sea water has been one of the most successful applications of membrane processes. The principle is that water is forced through a membrane which is almost impermeable to salt.

In reverse osmosis, tight but extremely thin polymer membranes are used. We shall treat the membrane as homogeneous. Concentrations in and around a typical reverse osmosis membrane are shown in Fig. 16.1. The distribution coefficients of water and salt differ considerably. That of water is about one-tenth, that of salt around one-hundredth. The concentration gradient of water in the membrane is relatively small; that of salt is considerable.

Before looking at the transport equations, we first analyse one special case. There is a pressure difference — the osmotic pressure difference — at which the driving force on water is zero. In our case the osmotic pressure has a value of 2.4 MPa. A reverse osmosis membrane only works well if the pressure difference is larger than the osmotic pressure.

In the transport relation for water (Fig. 16.2) only membrane friction is taken into account. The main driving force is the pressure gradient. The composition term is conventionally replaced by an osmotic pressure term: the water velocity is seen to be proportional to

reverse osmosis'

concentrations (mol l⁻¹)

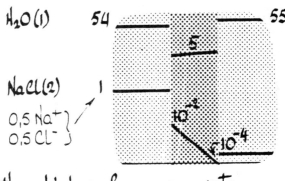

$H_2O (l)$ 54 55

5

$NaCl(2)$ 1

0,5 Na^+
0,5 Cl^-

10^{-4}

$\sim 10^{-4}$

the driving force on water
is zero when:

$$\frac{\Delta x_i'}{\overline{x}_i'} + \frac{\Delta \pi_1}{(140 \, MPa)} = 0 \longrightarrow \Delta \pi_1 = 2.4 \, MPa$$

osmotic pressure

there is only transport to the
right if $\Delta p \gtrsim \Delta \pi_1$

Fig. 16.1

the difference between the pressure difference and the osmotic
pressure difference.

In the salt transport equation the major driving force is the
composition term. The driving force on the salt is much larger than
that on the water. Even so, the salt velocity is lower. This is because
of the very much lower diffusivity of the hydrated salt ions.

The flux relations are summarized in Fig. 16.3. The water flux is
proportional to the pressure difference; the salt flux is barely dependent on it. So an increasing pressure differential improves the salt
retention.

In this example we have kept the equations very simple. Details
we have neglected in the calculations are:

- non-idealities,
- water–salt friction in the membrane,
- viscous flow and

reverse osmosis[2]

transport of water

$$\frac{\Delta x_1'}{x_1'} + \frac{\Delta p}{p_1^*} = \frac{\Delta p - \Delta \pi_1}{p_1^*} = -\frac{\bar{v}_1}{k_1}$$

$$+0.02 \quad -0.08$$

$$k_1 = \mathcal{D}_1 / \delta \approx 10^{-3} \, ms^{-1}$$

$$p_1^* = 140 \; MPa$$

$$\Delta p = 10 \; MPa \quad \longrightarrow \quad \bar{v}_1 = 5.4 \times 10^{-5} \, ms^{-1}$$

$$\Delta \pi_1 = 2.4 \; MPa \qquad N_1 = 0.3 \; mol \, m^{-2} s^{-1}$$

transport of salt

$$\frac{\Delta x_2'}{x_2'} + \frac{\Delta p}{p_2^*} = -\frac{\bar{v}_2}{k_2}$$

$$-2 \quad -0.33 \;\Bigg\} \quad \bar{v}_2 \approx 2 \times 10^{-5} \, ms^{-1}$$

$$k_2 \approx 10^{-5} \, ms^{-1} \Bigg\} \quad N_2 \approx 10^{-4} \, mol \, m^{-2} s^{-1}$$

Fig. 16.2

• the effect of the salt concentration downstream on the osmotic pressure difference.

If you have an equation solver on your computer you might try to include these effects. For realistic values of the parameters, the behaviour of the system is only slightly modified, except for pressure differences near the osmotic pressure.

ULTRAFILTRATION

Our second example is on the transport of a globular protein and water through a heterogeneous membrane. The ideas presented in this example are only rough approximations: we still have a lot to learn about ultrafiltration.

The membrane (Fig. 16.4) has cylindrical pores with a diameter of twenty nanometres. It is ten micrometres thick (the figure is not at all to scale). The pore cross-sections cover one tenth of the membrane area. Such membranes do exist, although they are not of great technical importance.

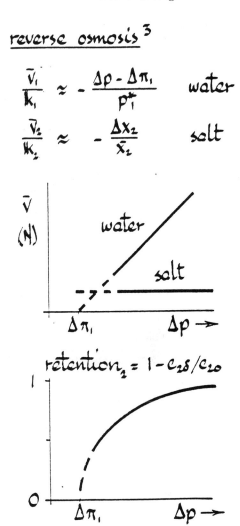

reverse osmosis [3]

$$\frac{\bar{v}_1}{k_1} \approx - \frac{\Delta p - \Delta \pi_1}{P_1^*} \quad \text{water}$$

$$\frac{\bar{v}_2}{k_2} \approx - \frac{\Delta x_2}{\bar{x}_2} \quad \text{salt}$$

retention$_2$ = $1 - c_{2s}/c_{2o}$

Fig. 16.3

The diameter of the protein is four nanometres. This is ten times larger than that of the solvent molecules (water). The volume fraction protein in the solution is low, so the molecules do not hinder each other. In the first part of this study we assume the protein to have no charge and no tendency to adsorb on the membrane.

Over the membrane there is a pressure drop of one hundred kPa. This causes both water and protein to move through the membrane. As we shall see, the separation achieved with this simple system is not large: the variation of the protein mole fraction shown in the figure is typical.

The step changes of the mole fraction protein at the pore inlet and outlet are due to geometrical exclusion (Fig. 16.5). The centre of the

ultrafiltration

a model membrane with cylindrical pores .. $\varepsilon \approx 0,1$

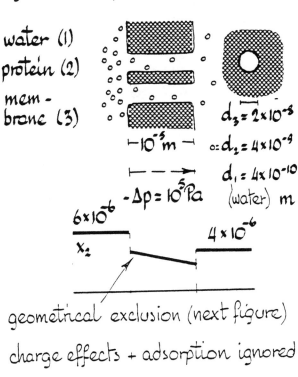

water (1)

protein (2)

mem-brane (3)

$d_3 = 2 \times 10^{-8}$

$\vdash 10^{-5} m \dashv$ $\circ : d_2 = 4 \times 10^{-9}$

$d_1 = 4 \times 10^{-10}$ (water) m

$-\Delta p = 10^5 Pa$

6×10^{-6}

4×10^{-6}

x_2

geometrical exclusion (next figure)

charge effects + adsorption ignored

Fig. 16.4

protein molecules cannot be within half a molecule diameter from the pore wall. When the diameter of the molecules approaches that of the pores this becomes important.

SLIPPING LAYER MODEL

Now we go on to the transport coefficients. As the tortuosity and constrictivity of our ideal membrane have a value of unity, the protein–water diffusivity is equal to that in free solution (Fig. 16.5). The transfer coefficient is obtained by dividing the diffusivity by the membrane thickness.

    Friction between water and the membrane occurs at the pore wall (Fig. 16.6). Here we assume that the friction is caused by water molecules slipping over the wall. The shear stress is described by the same formula as used for stresses in a viscous liquid. The only

geometrical exclusion
of the protein

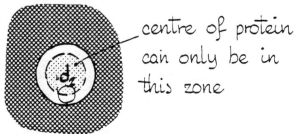

centre of protein
can only be in
this zone

$$\frac{x_1}{x_2'} = \frac{(d_3 - d_2)^2}{d_3^2}$$

$$= 0.64$$

water - protein friction

$r_{12} \leftarrow \bigcirc$ $\mathcal{D}_{12} \approx \mathcal{D}_{12}^{o} \approx 10^{-10} \, m^2 s^{-1}$

$\gamma = 1 \quad \tau = 1$ $k_{12} = \mathcal{D}_{12}/\delta \approx 10^{-3} \, m s^{-1}$

Fig. 16.5

difference is that the 'wall viscosity' need not be equal to the bulk liquid viscosity.

The friction between water and the pore wall is equal to the product of the shear stress and the interfacial area. From this friction it is a simple matter to obtain the expression for the water–membrane diffusivity given in the figure.

We know very little about wall viscosities. You might expect them to be higher than the bulk viscosity. They will also depend on the nature of the wall. Water should have a larger interaction with a hydrophilic polymer than with a hydrocarbon. The numbers given in the figure are for illustration only; the important point is that the water–membrane diffusivity is probably quite high here and that it increases with the pore diameter.†

SPHERE-IN-TUBE MODEL

To obtain the protein–membrane coefficient, we have calculated the friction of a sphere moving along the axis of a cylindrical tube (Fig.

---

† In reverse osmosis membranes the pore diameter is of the order of that of the water molecules. The sliding layer model then predicts a much larger friction, as is observed.

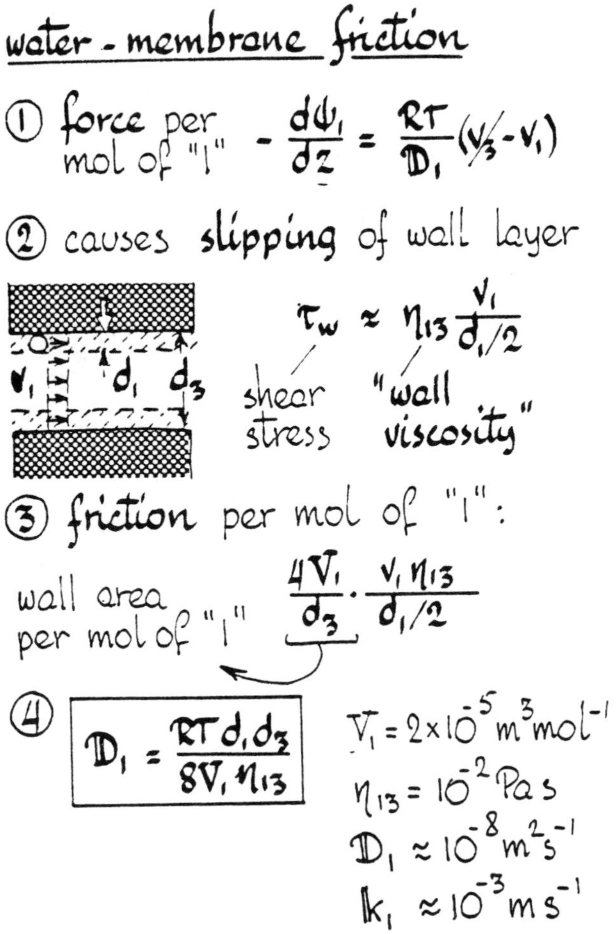

water - membrane friction

① force per mol of "1" $-\dfrac{d\psi_1}{dz} = \dfrac{RT}{D_1}(v_3 - v_1)$

② causes **slipping** of wall layer

$\tau_w \approx \eta_{13}\dfrac{v_1}{d_1/2}$

$v_1 \quad d_1 \quad d_3$ shear   "wall
stress   viscosity"

③ friction per mol of "1":

wall area per mol of "1" $\underbrace{\dfrac{4V_1}{d_3}}\cdot\dfrac{v_1\,\eta_{13}}{d_1/2}$

④ $\boxed{D_1 = \dfrac{RT\,d_1 d_3}{8V_1\,\eta_{13}}}$ $V_1 = 2\times 10^{-5}\,m^3 mol^{-1}$

$\eta_{13} = 10^{-2}\,Pa\,s$

$D_1 \approx 10^{-8}\,m^2 s^{-1}$

$k_1 \approx 10^{-3}\,m\,s^{-1}$

Fig. 16.6

16.7). The friction with the wall is then taken as the total friction minus that of the free sphere in an infinite liquid. The ratio of the protein–wall and protein–water transfer coefficients is the inverse of the friction coefficients. The protein–membrane friction increases very rapidly as the protein diameter approaches the pore diameter. For the diameter ratio in the example, the friction is not very large. There are probably better ways of making this estimate, but this is just as far as we had got when writing this book.

VISCOUS FLOW

In the relatively large pores used here, viscous flow must be taken into account (Fig. 16.8). The sliding layer model suggests that the effective tube diameter should be taken equal to the pore diameter minus the thickness of the wall layers. For the system here, this is unimportant, but it does suggest that viscous flow will disappear in very small pores.

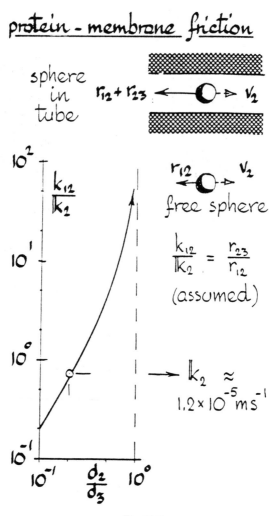

Fig. 16.7

Working out the transport relations (Fig. 16.9) shows a few interesting things. The viscous velocity is the dominant one. The diffusion velocity of water is negligible. That of the protein is not: in particular, the pressure gradient has some effect. (This is because of the very large molar volume of the protein.) So the protein moves through the pores slightly more rapidly than the water. The selectivity of the membrane towards water is caused solely by the geometrical exclusion of the protein.

Ultrafiltration is a complicated process and there are quite a few aspects we have not analysed (Fig. 16.10). To mention a few:

- polarization is very important in ultrafiltration. If you use large pressure differences the protein precipitates and forms a gel layer at the membrane interface,
- proteins often do adsorb on polymers. (The tendency to adsorb

viscous flow

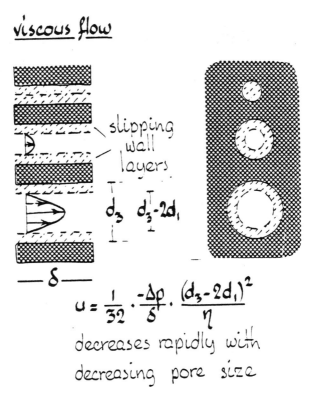

slipping wall layers

d_3 $d_3 - 2d_1$

$—\delta—$

$$U = \frac{1}{32} \cdot \frac{-\Delta p}{\delta} \cdot \frac{(d_3 - 2d_1)^2}{\eta}$$

decreases rapidly with decreasing pore size

Fig. 16.8

depends on the protein, the polymer interface, but also on the ionic strength and acidity of the solution.) Adsorption greatly increases the protein concentration in the membrane. It increases the protein–membrane friction even further and so improves the protein rejection.

● Both the proteins and the membrane often carry some charge. Like charges cause exclusion, opposite charges 'inclusion' and adsorption. The net result is not easy to predict.

SUMMARY (MEMBRANES)

In the last four chapters we have analysed six different membrane processes. We have seen that their performance is determined by many factors. They are:

● the driving forces (composition, electrical and pressure forces),
● friction between the mobile components,
● friction between the mobile components and the membrane,
● the component solubility which influences the concentration in the membrane and the flux calculations, and
● viscous flow.

viscous velocity

$$u \approx \frac{1}{32} \cdot \frac{-\Delta p}{\delta} \cdot \frac{d_3^2}{\eta} \approx 1.25 \times 10^{-4} \, m s^{-1}$$

diffusion velocities

water

$$\frac{\Delta x_1'}{\bar{x}_1'} + \frac{\Delta p}{p_1^*} = \bar{x}_2 \frac{\bar{v}_2 - \bar{v}_1}{k_{12}} + \frac{\bar{v}_3 - \bar{v}_1}{k_1}$$

$$2 \times 10^{-6} - 7 \times 10^{-4} \quad 4 \times 10^{-6} \quad 10^{-5} \quad 10^{-3}$$

$$\longrightarrow \bar{v}_1 = 0.7 \times 10^{-6} \, m s^{-1}$$

protein

$$\frac{\Delta x_2'}{\bar{x}_2'} + \frac{\Delta p}{p_2^*} = \bar{x}_1 \frac{\bar{v}_1 - \bar{v}_2}{k_{12}} + \frac{\bar{v}_3 - \bar{v}_2}{k_2}$$

$$-0.4 \quad -0.8 \quad 1 \quad 10^{-5} \quad 1.2 \times 10^{-5}$$

$$(p_2^* \approx 10^5 \, Pa!)$$

$$\longrightarrow \bar{v}_2 = 7 \times 10^{-6} \, m s^{-1}$$

Fig. 16.9

We have summarized which effects are important in Fig. 16.11. Large dots mean 'important', small dots 'of some importance', and no dot means 'not important'.

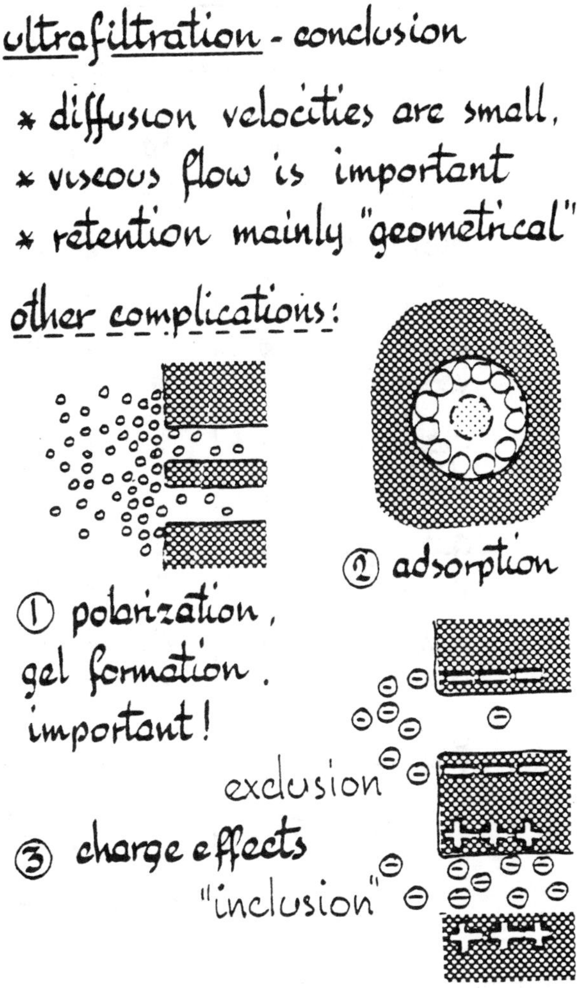

ultrafiltration - conclusion

* diffusion velocities are small,
* viscous flow is important
* retention mainly "geometrical"

other complications:

① polarization,
gel formation.
important!

② adsorption

exclusion

③ charge effects
    "inclusion"

Fig. 16.10

membrane processes

which terms are important?

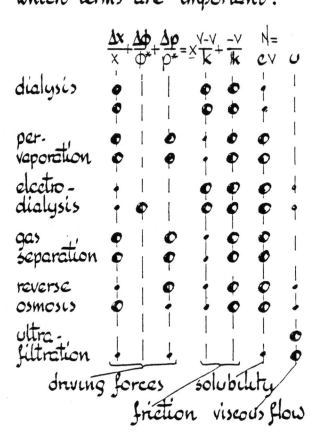

Fig. 16.11

17

Sorption

In the last regular chapter of this book, we take a simple view on two sorption processes: ion exchange and adsorption on activated carbon.

FIXED-BED PROCESSES

The important thing in a sorption process is the sorbent. It consists of permeable solid particles, which usually have a high affinity for the solute to be separated. The particle sizes range from tens of micrometres to several millimetres. There is an enormous variety of sorbents on the market.

Sorption processes are mostly carried out in a fixed bed of the sorbent particles (Fig. 17.1). Their prime application is for the

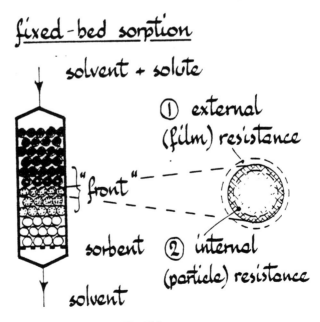

Fig. 17.1

removal of trace amounts of solutes from gas and liquid streams. The solvent and solute are pumped into a fresh column which contains little or no solute. As the solvent passes through the column, the solute is adsorbed. First the area near the feed point becomes saturated; this saturation zone gradually fills up the column. The sorbent usually stores the solute in a much higher concentration than that of the feed. So the column only fills up slowly.

After some time the 'front' reaches the end of the column and the sorption is stopped. The column is then regenerated. This is usually done by changing conditions in the column (pressure, temperature, electrical field, composition ...) in such a manner that the solute is no longer so strongly adsorbed. The solute is then removed in a concentrated form by passing a regenerant through the column.

The front is the actual mass transfer zone. Its width is quite an important design criterion. The front width increases with increasing column throughput and lower mass transfer capacities.

Just as in the other examples, we will restrict ourselves to mass transfer to and in a single particle. Also we will regard any mass transfer resistance as due to a 'film' in which steady-state transfer occurs.

There are two mass transfer resistances:

- the external (or film) resistance and
- the internal (or particle) resistance.

Both of these resistances usually play some role.

The external film is similar to the films we have seen earlier. Transport inside the particles is a very complicated process (Fig. 17.2). The solute first penetrates rapidly into the outer layers of the particle. After some time the fronts begin to overlap. In most cases a concentration profile develops of which the shape only changes a little. The largest part of the gradients is in the outer layers of the particle. Detailed calculations show that the resistance can then often be approximated by a diffusion 'film' with a thickness of around one tenth of the particle diameter.

This is all very approximate and you have some reason to be dissatisfied. However, a complete treatment of these transient multi-component problems is too difficult for this text, and you can learn quite a lot from the 'film' model.

As the diffusivities in the solid matrix are usually lower than those outside, you might expect the particle resistance to dominate. However, this is not necessarily so (Fig. 17.3). For a given flux of a species through the interface, the species velocity is highest in the phase with the low concentration. The solute concentration in the solvent is often orders of magnitude lower than in the sorbent, so the resistance in the solvent can be important.

ION EXCHANGE (FILM LIMITED)

In ion exchange a certain ionic species is removed from an aqueous solution and replaced by another. In the following examples the two

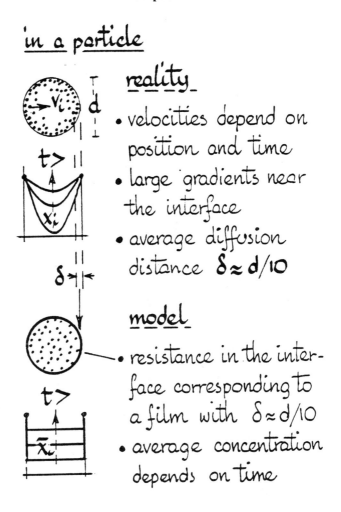

in a particle

reality
- velocities depend on position and time
- large gradients near the interface
- average diffusion distance $\delta \approx d/10$

model
- resistance in the interface corresponding to a film with $\delta \approx d/10$
- average concentration depends on time

Fig. 17.2

ions are from hydrogen and sodium. They are both positive. The ion exchanger particles are made of the same type of swollen polymer as electrodialysis membranes; here the particles have a negative charge. Again co-ions such as chloride are excluded from the matrix (at not too high external concentrations). Mass transfer in ion exchange is complicated by the transient nature of the process. This can cause swelling or shrinking of the particles. The resulting water flux through the interface can be taken into account by our models, but we shall not do so here.

We look at the film limitation outside the particle. This tends to be the overruling resistance at low external solute concentrations (in the sorption step).

In the example shown (Fig. 17.4) two situations are considered. In the first, sodium ions are removed from the electrolyte and replaced by hydrogen ions from the exchanger. In the second case the reverse

<u>dominant resistance</u>

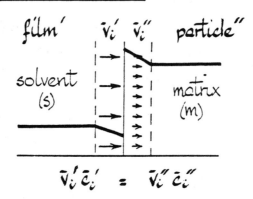

$$\bar{v}'_i \bar{c}'_i = \bar{v}''_i \bar{c}''_i$$

major friction terms usually:

$$r'_{is} = \underset{\approx 1}{\bar{x}_s} \frac{-\bar{v}'_i}{k'_{is}} \qquad \frac{\bar{v}''_i}{k''_i} = r''_{im}$$

$$\boxed{\frac{r'_{is}}{r''_{im}} = \frac{k''_i}{k'_{is}} \frac{c''_i}{c'_i}}$$

important for low external
                 concentrations

Fig. 17.3

occurs. In both situations the exchange has just begun and the
interface and bulk compositions have not yet changed appreciably.
Quite surprisingly these two processes are not at all symmetrical. The
second one is found to be considerably more rapid than the first.

The culprit is the hydrogen ion, in combination with the chloride
ion. The chloride does not participate in the exchange, but does
influence the mass transfer process. The fast-moving hydrogen ion
causes an electrical potential difference. In the left-hand example this
pushes chloride ions away from the interface. In the right-hand
example the reverse occurs. So in the left case there are low salt
concentrations and low gradients near the interface. In the right-hand
example it is merely the other way around.

A fair number of equations is required to describe these
examples. There are three transport relations for the three ions.
Additionally there are two bootstrap relations. One states that
chloride does not partake in the exchange. The other requires that
the net current be zero. Finally you need the electroneutrality

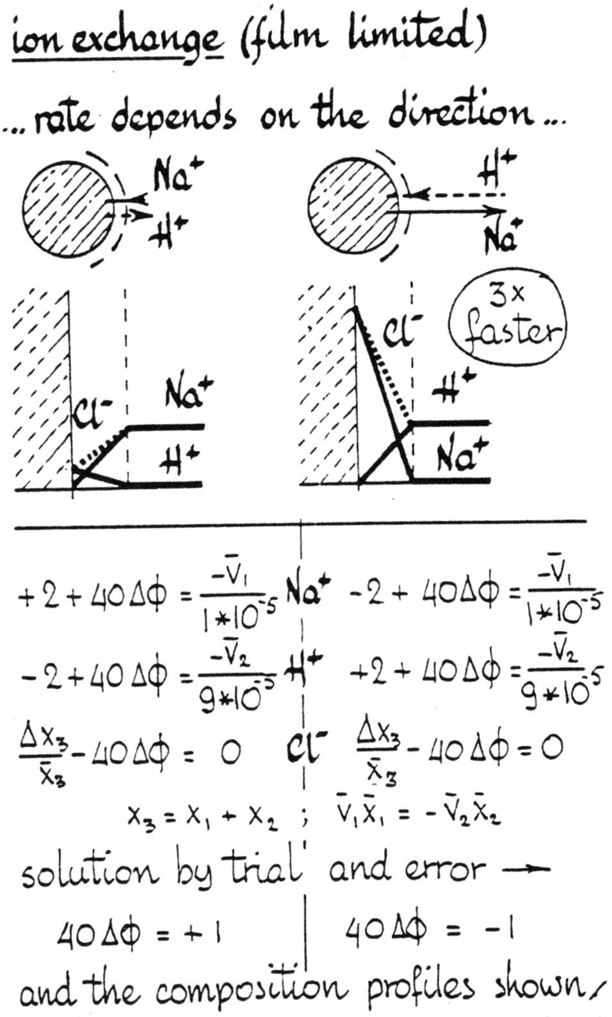

ion exchange (film limited)

... rate depends on the direction ...

$$+2 + 40\Delta\phi = \frac{-\bar{V}_1}{1*10^{-5}} \quad Na^+ \quad -2 + 40\Delta\phi = \frac{-\bar{V}_1}{1*10^{-5}}$$

$$-2 + 40\Delta\phi = \frac{-\bar{V}_2}{9*10^{-5}} \quad H^+ \quad +2 + 40\Delta\phi = \frac{-\bar{V}_2}{9*10^{-5}}$$

$$\frac{\Delta x_3}{\bar{x}_3} - 40\Delta\phi = 0 \quad Cl^- \quad \frac{\Delta x_3}{\bar{x}_3} - 40\Delta\phi = 0$$

$$x_3 = x_1 + x_2 \; ; \; \bar{V}_1\bar{x}_1 = -\bar{V}_2\bar{x}_2$$

solution by trial and error →

$$40\Delta\phi = +1 \quad | \quad 40\Delta\phi = -1$$

and the composition profiles shown,

Fig. 17.4

relation. The ionic mole fractions at the interface are not known beforehand. Therefore these equations have to be solved by trial and error (or with an equation solver).

In the partly worked-out example we have taken the mass transfer coefficient of hydrogen to be nine times that of sodium. (In reality the difference is not quite as large, but this gives nice round figures . . . .) You can easily check that the electrical terms in the left- and right-hand equations are then plus unity and minus unity respectively. The fluxes in the right-hand case turn out to be three times higher than in the left-hand case.

ION EXCHANGE (PARTICLE LIMITED)

With high external concentrations, the external mass transfer resistance may become unimportant. Transport is then limited by phenomena inside the particle (Fig. 17.5). We will regard the particle as homogeneous and again neglect swelling. We will also neglect any co-ion transport, although this is not necessarily correct at high external concentrations.

We consider the case that the exchanger has initially been in the hydrogen form. Only a trace of hydrogen is left; it is diffusing outward. From the 'no-current' relation we see that the sodium velocity is much lower than the hydrogen velocity. All terms in the sodium transport relation are then very small. This also applies to the electrical gradient. In the hydrogen relation we then see that the transport rate is completely determined by friction of the trace component with its surroundings. In this case it is by the faster of the two ions! For you this may be obvious by now. It was not so for earlier investigators, who were used to thinking in terms of the Fick description of diffusion.

As you can also see, at high external concentrations, ion exchange

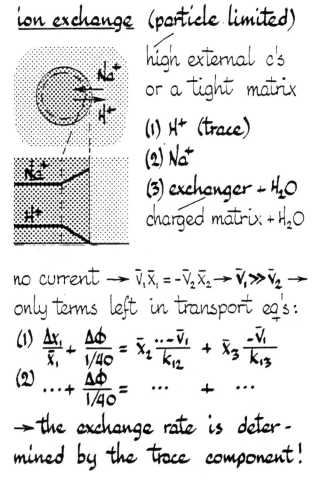

ion exchange (particle limited)

high external c's
or a tight matrix

(1) H^+ (trace)
(2) Na^+
(3) exchanger + H_2O
charged matrix + H_2O

no current → $\bar{v}_1 \bar{x}_1 = -\bar{v}_2 \bar{x}_2$ → $\bar{v}_1 \gg \bar{v}_2$ →
only terms left in transport eq's:

(1) $\dfrac{\Delta x_1}{\bar{x}_1} + \dfrac{\Delta \phi}{1/40} = \bar{x}_2 \dfrac{\cdots - \bar{v}_1}{k_{12}} + \bar{x}_3 \dfrac{-\bar{v}_1}{k_{13}}$

(2) $\cdots + \dfrac{\Delta \phi}{1/40} = \cdots + \cdots$

→ the exchange rate is determined by the trace component!

Fig. 17.5

rates depend on the exchange direction. Here we have considered the last part of the exchange. A little thought will show you that the effect is just the opposite in the beginning of the exchange.

ADSORPTION

In solid sorbents, large surfaces are required to obtain a reasonable capacity. This implies very narrow (micro)pores, in which diffusion is slow. To speed things up, sorbents are made as agglomerates of fine sub-particles. The diffusion to the sub-particles occurs through the macropores between them (Fig. 17.6). With such agglomerates three main diffusional resistances can be distinguished:

- the film resistance outside the particle,
- the macropore resistance and
- the micropore resistance.

In a rough kind of way you can regard these three as resistances in series.

To keep something real in mind we use an example: adsorption of

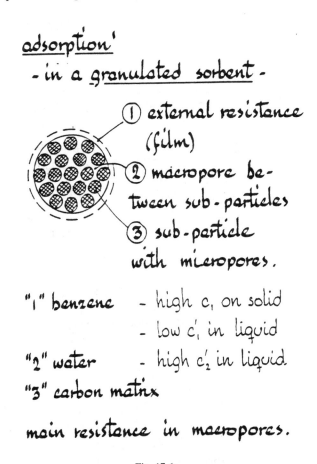

adsorption'
- in a granulated sorbent -

① external resistance (film)

② macropore between sub-particles

③ sub-particle with micropores.

"1" benzene   – high c_1 on solid
              – low c_1' in liquid

"2" water   – high c_2' in liquid

"3" carbon matrix

main resistance in macropores.

Fig. 17.6

benzene from water on activated carbon. Here the micropores have a diameter of around one nanometre; the macropores are up to a hundred times wider. Their length is much greater than their width. For a small molecule such as benzene, mass transfer is mainly limited by macropore diffusion.

The benzene concentration in the water is very low. However, benzene adsorbs preferentially on the carbon and its concentration there can be very much higher (Fig. 17.7). Intuitively, you would expect the transport to be due to diffusion of the free benzene in the pore liquid. Experiment, however, shows that the rates are usually more than one order of magnitude higher than can be explained by this mechanism. The main way transport occurs appears to be by surface diffusion. The low surface mobility is more than compensated by the higher concentration.

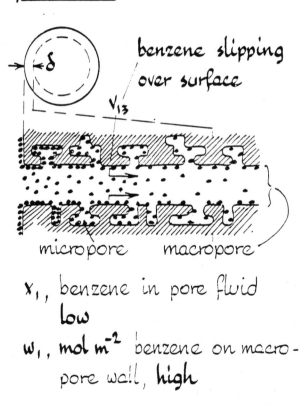

pore model

benzene slipping over surface

v_{13}

micropore　　　macropore

x_1, benzene in pore fluid low

w_1, mol m^{-2} benzene on macro-pore wall, high

Fig. 17.7

A few aspects of the system are worked out in Fig. 17.8. The potential of benzene in the pore liquid is given by the usual formula for an ideal mixture. The wall is assumed to be in equilibrium with that liquid. The relation between the wall and liquid concentrations can often be approximated by a simple power law, in which the

potentials

in pore fluid: $\psi_1 = RT \ln x_1$

equilibrium between pore fluid and wall:

$$w_1 = \mp x_1^n$$

$\longrightarrow$ also $\psi_1 = \dfrac{RT}{n} \ln w_1 + \text{const}'$

benzene velocity (over surface)

$$V_{1s} = -\frac{D_1}{RT}\frac{d\psi_1}{dz} = -\frac{D_1}{n w_1}\frac{dw_1}{dz}$$

assuming major friction is between benzene and surface

flux

$$N_1 = \overline{V}_1 \overline{w}_1 a = -\frac{D_1}{\delta}\frac{a}{n}\Delta w_1$$

macropore surface, $m^2 m^{-3}$

Fig. 17.8

exponent is considerably smaller than one. Also the friction between the benzene and the surface is thought to be dominating. It is then a simple matter to derive that the flux into the particle is proportional to the gradient of the surface concentration.

We still know rather little about Maxwell–Stefan surface diffusivities. In systems like this they are one or two orders of magnitude lower than free diffusivities, and they appear to be fairly independent of surface concentration.

In this example we only had one component strongly adsorbing. With more sorbates on the pore wall you must expect interactions, but we shall not pursue this subject here.

SUMMARY

- Sorption is always a transient process. Simple 'film' resistances describe such a process only approximately.

- There are mass transfer resistances both inside and outside the particle. The external resistance can be important, especially with low external concentrations.
- In ion exchange, mass transfer rates depend on the direction of transfer. We have analysed the situation both for external and for internal resistances.
- In a heterogeneous sorbent such as activated carbon, the major transport mechanism is often surface diffusion.

18

Alternatives

A complicated subject such as multicomponent mass transfer can be described in many ways. We have focused here on one type of description: the 'driving force = friction' or Maxwell–Stefan-type of approach. You may wonder why. We will tell you using an example. The example is diffusion in an ideal ternary mixture, where diffusion is caused by concentration gradients only.

THREE WAYS

Three common ways of describing this system are shown in Fig. 18.1. The first should by now be familiar to you, even though it is in the form of differential equations (and not difference equations). The second is the form mostly used in the thermodynamics of irreversible processes (TIP). The third shows the Fick equations, which make the process engineer feel at home. There are many variations possible on these equations, for example in the units used. All this is fairly obvious.

The three descriptions — although they look very different — can be transformed into each other. They give the same results and are all equally correct. So you choose the one you find most handy or the easiest to understand. Also in all three systems you must choose any two out of the three transport relations for the three components.

THE FICK DESCRIPTION

In the first instance the Fick equations look attractive. They seem a logical extension of mass transfer theory, and look the right way for a process engineer. Also the fluxes are explicit and concentrations are handy in calculations involving mass balances. No thermodynamic model is required in mass transfer calculations. There are, however, disadvantages.

three ternary alternatives

① Maxwell - Stefan (MS)

$$\frac{1}{RT}\frac{d\mu_1}{dz} = x_2 \frac{v_2 - v_1}{Ð_{12}} + x_3 \frac{v_3 - v_1}{Ð_{13}}$$

$$\frac{1}{RT}\frac{d\mu_2}{dz} = x_1 \frac{v_1 - v_2}{Ð_{12}} + x_3 \frac{v_3 - v_2}{Ð_{23}}$$

② TIP

$$J_1 = L_{11} Y_1 + L_{12} Y_2$$

$$J_2 = L_{12} Y_1 + L_{22} Y_2$$

$$Y_1 = -\left(1 + \frac{x_1}{x_3}\right)\frac{d\mu_1}{dz} - \left(\frac{x_2}{x_3}\right)\frac{d\mu_2}{dz}$$

$$Y_2 = -\left(\frac{x_1}{x_3}\right)\frac{d\mu_1}{dz} - \left(1 + \frac{x_2}{x_3}\right)\frac{d\mu_2}{dz}$$

③ Fick

$$J_1 = -cÐ_{11}\frac{dx_1}{dz} - cÐ_{12}\frac{dx_2}{dz}$$

$$J_2 = -cÐ_{21}\frac{dx_1}{dz} - cÐ_{22}\frac{dx_2}{dz}$$

Fig. 18.1

The driving forces in the Fick equations are concentration gradients. If you try to include the effects of non-idealities and electrical and pressure gradients the equations become very messy and almost incomprehensible. Another problem is the behaviour of the diffusivities. In a binary gas mixture (see Chapter 2) the diffusivity is independent of composition. You might expect such simple behaviour also to prevail in ternary mixtures. This is not so.

As an example we have taken the diffusivities in a ternary mixture of ideal gases. (This is the only kind of system in which we can calculate ternary diffusivities with any accuracy.) The pressure is 100 kPa and the temperature 298 K. The components are hydrogen, nitrogen and di-chloro-di-fluoro-methane (a 'Freon'). The molar masses of the three gases have been chosen far apart to accentuate a few things, but otherwise the system is nothing special.

You must choose two independent components and this can be done in three ways (Fig. 18.2). In the first we have chosen nitrogen

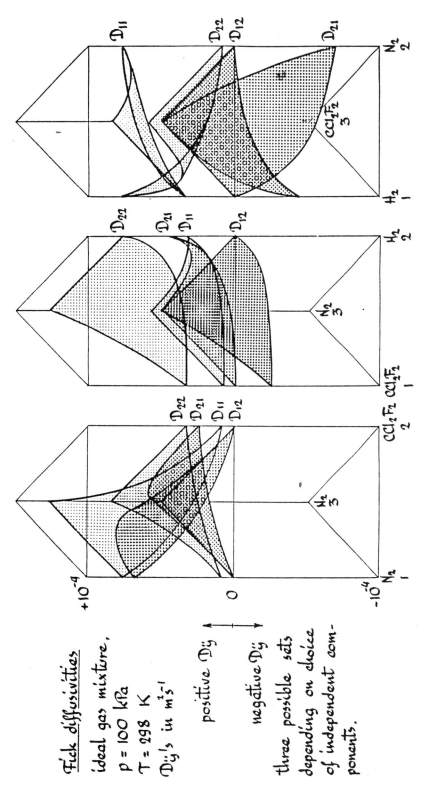

Fig. 18.2

and the Freon. The four diffusivities are plotted in perspective above a triangular composition diagram. You see that all four coefficients are positive. However, their behaviour is far from simple: they form strongly curved surfaces in space. Only two straight lines are visible along the binary boundaries.

If you choose the Freon and nitrogen as the independent components (second figure) the behaviour is completely different. In this case one of the cross-coefficients is always negative. Clearly the size of the diffusivities depends not only on pressure, temperature and composition, but also on the choice of the sequence of the components!

The result of the third choice is even nicer and wilder. The two cross-coefficients are now both negative, and see how beautifully curved in space the two main coefficients are.

We have never tried this on a non-ideal system, but strongly suspect that it will yield a scientific edifice of great and lasting proportions. This is nice for 3D-graphics, but not for your comprehension.

THERMODYNAMICS OF IRREVERSIBLE PROCESSES

The TIP representation can be derived in a straightforward manner from fairly simple assumptions on entropy production. It is explicit in the fluxes (which remains an advantage in numerical calculations). Also the form is such that you can immediately see what the effects of changing the driving forces are. The driving forces are easily modified to incorporate non-idealities, electrical and pressure gradients, etc. Here the 'cross-coefficients' are equal, so only three coefficients are required.

However the complicated behaviour of the coefficients remains. Also the behaviour depends on which components are chosen as independent (Fig. 18.3). The difference approximation in the TIP description usually has a smaller range of validity than that of the following description.

THE MAXWELL–STEFAN DESCRIPTION

You already know the behaviour of the Maxwell–Stefan diffusivities (Fig. 18.4). There are just three of them. They are independent of composition and the same for the three choices of independent components. The Maxwell–Stefan equations are thought to be difficult to use. We hope to have shown you that this is not at all so if you accept a few approximations.

The pros and cons of the three systems are summarized in Fig. 18.5.

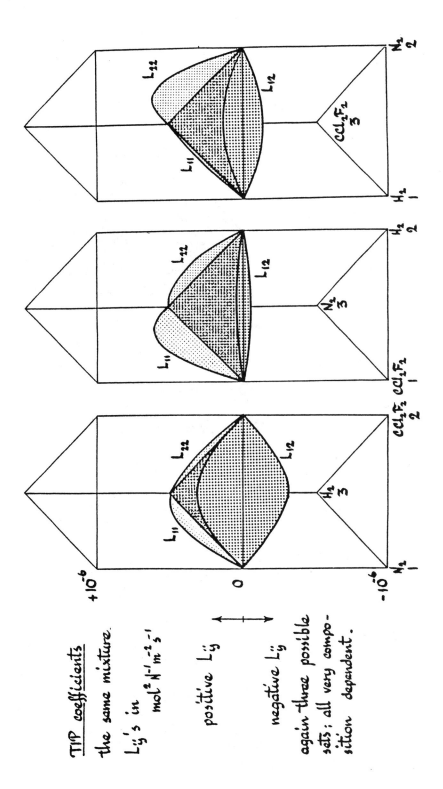

Fig. 18.3

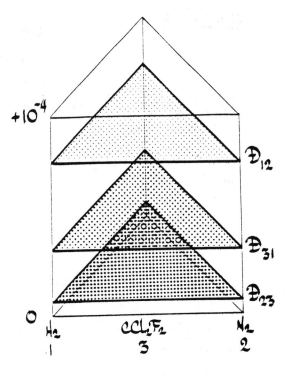

$+10^{-4}$

$\mathcal{D}_{12}$

$\mathcal{D}_{31}$

$\mathcal{D}_{23}$

O

H_2 1 CCl_2F_2 3 N_2 2

MS - diffusivities

this does the same job as the previous sets.

Fig. 18.4

advantages

	MS TIP	Fick	
"simple" behaviour of transport coeff's.	+	-	-
independent of reference frame	+	-	-
easily extended to other driving forces	+	+	-
number of transport coefficients (ternary)	3	3	4
transport coeff's independent of component sequence	+	-	-
fluxes explicit	-	+	+
easily integrated with thermodynamics	+	+	-
looks like classical chemical engineering	-	-	+

Fig. 18.5

19

. . . Ending

This has been a short — and possibly unconventional — tour through the vast subject of mass transfer. In contrast with most other engineering texts we have considered:

- more than two components and
- driving forces other than concentration gradients.

We hope that you feel that this is a worthwhile extension. You have seen that all mass transfer processes can be understood from a single viewpoint. And you now have a basis for describing and understanding the new separation processes that are the challenge for this generation of engineers.

Now you have had a start we hope you would like to learn a bit more on this subject. There is a list of further readings below. You will find an enormous amount of information there. There is also much that you will not find. Multicomponent mass transfer is still in its infancy. It is your turn now.

FURTHER READING

The literature on mass transfer is far too extensive for us to even try to be complete. So we have made a selection, which is of course coloured by our preferences. The books are chosen either because they treat mass transfer along the Maxwell–Stefan (MS) lines, or because they have something special. The articles contain additional information on subjects treated in this text.

BOOKS

[1] R. B. Bird, W. E. Stewart and E. N. Lightfoot, *Transport Phenomena*, Wiley, New York, 1960.
This book contains the definitive treatment of binary mass

transfer. Drift phenomena are covered very clearly. It also gives some information on binary transport coefficients in different systems. The part on multicomponent mass transfer is a little dated.

[2] E. N. Lightfoot, *Transport Phenomena and Living Systems*, Wiley, New York, 1974.

One of our favourites. It contains a clear introduction to the MS approach of multicomponent mass transfer. Also a nice set of examples on electrochemical systems and a very good treatment of membrane processes.

[3] R. Krishna and R. Taylor, Chapter 7 in *Handbook for Heat and Mass Transfer Operations*, Vol II, N. P. Cheremisinoff (editor), Gulf Publishing Corporation, Houston, 1986.

A rigorous treatment of multicomponent mass transfer along the M–S approach. Many examples on multicomponent condensation, distillation and liquid–liquid extraction. Good coverage of simultaneous heat and mass transfer. Extensive references.

[4] R. Krishna and R. Taylor, *Multicomponent Mass Transfer*, to be published by Wiley, New York, in 1990 or 1991.

A comprehensive treatment of multicomponent mass transfer along the M–S line. Many fully worked examples. Covers the numerical problems in simulations of chemical engineering systems. Also very extensive references.

[5] E. L. Cussler, *Diffusion, Mass Transfer in Fluid Systems*, Cambridge University Press, Cambridge, 1984.

Diffusion in the neighbourhood of the critical point.

[6] R. C. Reid, J. M. Prausnitz and B. E. Poling, *The Properties of Gases and Liquids*, 4th Edition, McGraw-Hill, New York, 1988.

The standard text on the determination of diffusivities in gases and liquids. Contains little information on multicomponent systems.

[7] J. O'M. Bockris and A. K. N. Reddy, *Modern Electrochemistry I*, Plenum Press, New York, 1970.

A lengthy, but very readable introduction to electrochemical systems. Only a cursory introduction to the frictional approach to mass transfer.

[8] J. Newman, *Electrochemical Systems*, Prentice-Hall, Englewood Cliffs, N.J., 1973.

An excellent book, with an extensive treatment of multicomponent mass transfer in electrolytes along the M–S approach.

[9] K. S. Forland, T.Forland and S. K. Ratkje, *Irreversible Thermodynamics, Theory and Applications*, Wiley, New York, 1988.

Many examples from electrochemistry, using the equations of the thermodynamics of irreversible processes.

156 Ending [Ch. 19

[10] R. Jackson, *Transport in Porous Catalysts*, Elsevier, Amsterdam, 1977.

[11] E. A. Mason and A. P. Malinauskas, *Gas Transport in Porous Media: the Dusty Gas Model*, Elsevier, Amsterdam, 1983.
[10] and [11] are good books on the diffusion of gases in porous media.

[12] K. S. Spiegler, *Principles of Energetics*, Springer-Verlag, Berlin, 1983.
An introductory text, containing many examples of different approaches to mass transfer processes. Nice examples on membranes.

[13] F. Helfferich, *Ion Exchange*, McGraw-Hill, New York, 1962.
Probably still the best introduction to mass transfer in charged matrices and in ion exchange.

[14] D. M. Ruthven, *Principles of Adsorption and Adsorption Processes*, Wiley, New York, 1984.
A recent, well-known work on adsorption. The frictional approach is applied, but mainly to adsorption of a single component.

ARTICLES

These are arranged by chapter.

Chapter 2. The two-bulb ternary gas experiment is described by
J. B. Duncan and H. L. Toor, *AIChE J*, **8**, 38 (1962).
Information on charged membranes and two-gases-with-a-plug is to be found in the books [11] and [13].

Chapters 3 and 4. A masterly discussion on the models for interphase mass transfer, with emphasis on the hydrodynamics close to the interface is to be found in
L. E. Scriven, *Chem. Eng. Educ.*, Autumn 1968, pp. 150–155; Winter 1969, pp. 26–29; Spring 1969, pp. 94 98.
The sphere-in-liquid model is from A. Einstein; an English translation of his relevant articles is to be found in
A. Einstein, *Investigations on the Theory of Brownian Movement*, Dover Publications Incorporated, New York, 1956.
Einstein uses the gradient of the osmotic pressure as the driving force. Otherwise his treatment is quite in line with the M–S approach.

Clear descriptions of the various driving forces causing mass transfer, and derivations of the friction formulation are given in [2] and [12]. More formal treatments emphasizing diffusion in multicomponent mixtures are available in [3] and [4].

Chapter 5. Binary mass transfer has been the subject of scores of textbooks. However, not all authors have paid sufficient attention to drift and the various bootstrap relations. The classic and definitive treatment is the book of Bird *et al.* [1].

Chapter 6. See [3] and [4].

Chapter 7. See [3] and [4]. The example on absorption of ammonia in a packed column is from
J. D. Raal and M. K. Khurana, *Can. J. Chem. Eng.*, **51**, 162 (1973).

Chapter 8. Thermodynamics and diffusion near critical points and in non-ideal mixtures are treated in [4] and [5].

Chapter 9. The best text on estimation of diffusion coefficients in gaseous and liquid mixtures is [6]. The description of the concentration dependence of the M–S diffusivity is due to
A. Vignes, *Ind. Eng. Chem. Fundamentals*, **5**, 189 (1966).
The estimation of mass transfer coefficients for various hydro-dynamic situations is discussed in [1].

Chapter 10. Other forces are the subject of [2] and [12].

Chapter 11. Excellent treatments of diffusion in electrolytes are available in [2] and [8]. Mixed ion diffusion effects were first high-lighted by
J. Vinograd and J. W. McBain, *J. Am. Chem. Soc.*, **63**, 2008 (1941).
Chapters 12–16. Good treatments of membrane processes using the M–S approach are to be found in [2] and [12]. The dusty gas model is described in [10] and [11].

An interesting study on electrodialysis is
E. M. Scattergood and E. N. Lightfoot, *Trans. Faraday Soc.*, **64**, 1135 (1968).

Chapter 17. For ion exchange and adsorption see [13] and [14].

Chapter 18. See [3], [4] and [12].

Thanks

Many people have contributed to this book in some way or other. We would like to thank them. The first ones are the students on two graduate courses on multicomponent mass transfer that we gave at Delft University of Technology in 1983 and 1984. The problems we encountered trying to teach them led to a series of eight short articles 'Beweging in Mengsels' (movement in mixtures). These appeared in our Dutch chemical engineering magazine *I2-Procestechnologie* in 1986–87. They may be regarded as the forerunners of this book.

Of the material used, the data on ion–ion interactions and the external film model in ion exchange were sorted out by Margot Baerken. The TIP coefficients in Chapter 18 were calculated by Pieter Vonk. Both of them were students in Delft. The pervaporation example is from Wim Groot, of the Laboratory of Biotechnology of Delft University. The idea of the activated carbon–benzene–water example model originated in discussions with Wim van Lier of NORIT N.V. (activated carbon), Amersfoort. On the membrane examples we had many discussions with Joop Bitter of Shell Research Amsterdam and Jan-Henk Hanemaayer of TNO, Zeist. And we must thank Ross Taylor of Clarkson University, Potsdam, in the United States. He carefully checked all figures in the original version of this book. Ross also played a prominent role in our third course in 1988. Daniel Tondeur of the CNRS in Nancy, France, gave extensive and very helpful comments.

A man who may not realize that he had an impact on this book is Alirio Rodrigues from Porto, Portugal. He was the main organizer of a series of NATO summer schools in Portugal. Courses on reaction engineering, ion exchange and adsorption brought us in contact with people such as Friedrich Helfferich, Gerhard Klein, Douglas Ruthven and Patrick Meares. They all contributed to our thinking about the subject. The summer course in 1988 also gave Hans the opportunity to start writing this book. The first version was finished in the home of Theresa and Leo Mudde in Lissabon, during the hot summer afternoons, in a large dusky room, shuttered to keep the sun out.

The final writing took place during 1988–89. The Koninklijke Nederlandse Academie voor Wetenschappen and the Shell group of companies provided financial support for a sabbatical year for Krish. He spent this with Professor Ton Beenackers at the University of Groningen, whom he would very much like to thank for the congenial atmosphere. This allowed us to prepare and give another one week's intensive course in 1989. There we tried out the book on forty-four participants, mainly research students from the whole of the Netherlands. They discovered a fair number of the errors in the text, and got us to set the exercises accompanying this text.

The one who probably suffered most from the writing of the book is Trudi Wesselingh, but she did not complain for a moment.

We must have forgotten others, but hope they will forgive us.

Hans Wesselingh/Rajamani Krishna
Groningen/Amsterdam, July 1990

Symbols

Symbols which are only used at a single point in the text are not listed.

Symbol	Description	Value	Units
A	non ideality parameter		–
c	total (mixture) concentration		$mol\ m^{-3}$
c_i	concentration of species i		$mol\ m^{-3}$
c_p	molar heat capacity		$J\,mol^{-1}K^{-1}$
d	diameter		m
D	Fick diffusivity		$m^2\,s^{-1}$
$Đ$	Maxwell-Stefan diffusivity		$m^2\,s^{-1}$
$\mathbb{D}$	$= Đ_{im}/x_m$, membrane diffusivity		$m^2\,s^{-1}$
E	energy flux		$W\ m^{-2}$
f_i	driving force on i		$N\,mol^{-1}$
F	Faraday constant	96.4×10^3	$C\,mol^{-1}$
g	gravitational acceleration	9.81	$N\,kg^{-1}$
h	heat transfer coefficient		$W\ m^{-2}K^{-1}$
H	enthalpy		$J\ mol^{-1}$
J_i	flux with respect to mixture		$mol\ m^{-2}s^{-1}$
k	mass transfer coefficient		$m\,s^{-1}$
$\mathbb{k}$	$= k_{im}/x_m$, membrane coefficient		$m\,s^{-1}$
L	Avogadro number	6.02×10^{23}	$number\ mol^{-1}$
M_i	molar mass of i		$kg\ mol^{-1}$

N_i	flux with respect to interface		$mol\ m^{-2}\ s^{-1}$
p	(total) pressure		$N\ m^{-2}$
p_i	partial pressure of i		$N\ m^{-2}$
p_i^*	$= RT/V_i$, constant in driving force		$N\ m^{-2}$
r_i	frictional force on i		$N\ mol^{-1}$
r_i^*	dimensionless centrifugal potential		$-$
R	gas constant	8.314	$J\ mol^{-1}\ K^{-1}$
T	absolute temperature		K
v	viscous flow velocity		$m\ s^{-1}$
v_i	species velocity of i		$m\ s^{-1}$
V_i	species ($\equiv$ partial molar) volume		$m^3\ mol^{-1}$
x_i	mole fraction of i		$-$
z	distance		m
z_i	charge number of i		$-$
γ_i	activity coefficient of i		$-$
δ	film thickness		m
Δ	increment, i.e. $\Delta x_i = x_{i\delta} - x_{io}$		$-$
ε	void fraction		$-$
ε_i	volume fraction of i		$-$
η	viscosity		$N\ s\ m^{-2}$
μ_i	chemical potential of i		$J\ mol^{-1}$
ς	density		$kg\ m^{-3}$
σ	volume in diffusion correlations		$m^3\ mol^{-1}$
ϕ	electrical potential		V
ϕ_i^*	$= RT/(Fz_i)$, constant in driving force		V
ψ_i	(total) potential of i		$J\ mol^{-1}$
ω	rotational frequency		s^{-1}

subscripts

o δ left, right side of film

i component under consideration

j another component

m matrix, membrane

p particle, pore

w wall, water

1 2 component 1, component 2

12 , interaction between "1" and "2"

$+$ $-$ positive, negative ions

superscripts

$-$ taken at the average composition
in a film : $\bar{x}_i = (x_{io} + x_{i\delta})/2$
$$\bar{v}_i = v_i \text{ at } x_i = \bar{x}_i$$

$x_i = 1$ in pure i

$*$ in driving force constants

$'$ in the other phase; outside the membrane.

o in the free solution

Worked exercises

INTRODUCTION

This is a set of exercises accompanying the text of MASS
TRANSFER. They are meant as exercises for the reader, and
originally we did not plan to provide solutions. However, our
students have persuaded us to do so. Nevertheless, you will
learn more by first trying them yourself.

A few remarks about the exercises:
(1) They differ greatly in size and the amount of work
required. The average time might be one hour, but there are
also a few on which you might spend a day.
(2) Many of the exercises require you to to determine
velocities or fluxes through a film. Broadly speaking there
are two greatly differing levels of complexity:
- Exercises in which the driving forces are specified. This
usually means that you know the compositions on both sides of
the film beforehand. These exercises lead to small sets of
linear equations. These are easily solved.
- If the determination of the forces is part of the exercise,
life is much more difficult. The equations are usually non-
linear and there may be many more of them. Finding which
equations are required to define the problem may require some
thinking. Also immediate convergence to the solution is not
assured. We often find that these problems need to ripen
overnight!

MathCAD

All exercises have been worked out using MathCAD 2.5. This is
a program which is described by the manufacturer as an
electronic scratchpad. It is available from MathSoft Inc.,
201 Broadway, Cambridge, Massachusetts 02139, USA.
There are versions for PCs and Macintosh computers.

MathCAD freely mixes text, equations and plots. The notation
is similar to standard mathematics. You should be able to
follow nearly all of the reporting without any knowledge of
the program. A few points which may not be obvious are shown
in the examples below.

MathCAD can solve sets of linear and non-linear equations.
One example: determining the point of intersection of a
linear and a cubic equation.

$x := 1 \qquad y := 1$ define the variables and set
                              initial values.

Given   start of the solve block

$y \approx 2 \cdot x^3 - 4 \cdot x^2 + 3 \cdot x + 1$ equations to be solved
$y \approx 3 \cdot x$

$\begin{bmatrix} x \\ y \end{bmatrix} := \text{Find}(x,y)$ solves the variables x and y
                              and stores them in a vector

$$\begin{bmatrix} x \\ y \end{bmatrix} = \begin{bmatrix} 0.597 \\ 1.791 \end{bmatrix}$$ shows the resulting vector

If the equations are non-linear, it may be important to have good initial values. With a poor choice you may converge to a wrong answer, or not converge at all.

Plotting is done in MathCAD using range variables.

i := 0 ..100   range variable

$$x_i := \frac{i}{50}$$ the ordinate

$y(x) := 2 \cdot x^3 - 4 \cdot x^2 + 3 \cdot x + 1$

$z(x) := 3 \cdot x$ functions to be plotted

Units

For typographical reasons the units are not given in most examples. Except where mentioned, all units are standard SI values, without decimal prefixes.

This should be sufficient to allow you to follow the problem-solving method in the different examples.

1. FORCES IN A GLASS OF BEER

The question in this exercise is given in Fig. 3.7. You are to calculate the force on the carbon dioxide in a glass of beer.

Solution

We need the following parameters:

$xo := 0.003 \quad x\delta := 0.001 \quad \delta := 1 \cdot 10^{-5} \quad R := 8.314 \quad$ and

$T := 280$ (a decent temperature for beer). This yields:

$$f := -\left[\frac{R \cdot T}{\delta \cdot 0.5}\right] \cdot \frac{x\delta - xo}{x\delta + xo} \qquad f = 2.328 \cdot 10^{8}$$

The units are N/mol. So this is a proper driving force. To obtain the force in kgf/kg you must divide by the weight of one mole:

$M := 44 \cdot 10^{-3} \qquad g := 9.81$

$f' := \dfrac{f}{M \cdot g} \qquad f' = 5.393 \cdot 10^{8}$ \qquad Indeed, a force of more than five hundred thousand tons!

2. GRAVITATIONAL VERSUS CHEMICAL POTENTIAL

Chemical potential gradients yield forces which are of the order:

$$F_\mu \approx 10^8 \quad N/mol$$

You will find similar values in different situations, independent of the particle size. This is quite different for the potential gradient due to gravity (Figure 3-4). Estimate for which particle size gravity will become more important than the chemical potential gradient.

Solution.

The force due to gravity is: $\quad F_g := -M \cdot g$

with $\qquad\qquad\qquad M \approx \rho \cdot d^3 \cdot L$

Here g is the gravitational acceleration, ρ the density, d the particle diameter and L is Avogadro's number. With

$$g := 10 \quad \rho := 10^3 \quad L := 6 \cdot 10^{23}$$

the two forces are seen to be of the same order of magnitude when:

$$d := \left[\frac{10^8}{\rho \cdot L}\right]^{0.3333} \quad \text{or} \quad d = 6 \cdot 10^{-7}$$

For particles smaller than this, the effect of gravity diminishes rapidly. For much larger particles, chemical potential gradients become unimportant. For particles of around one micrometre diameter, both sedimentation (gravity) and diffusion (chemical potential) effects play a role.

3. DOES A COLD FLOOR CREATE OXYGEN?

In the winter, the floor of a room can be appreciably colder than the ceiling. A difference in temperature of 15 C is not exceptional. According to the ideal gas law this implies that the concentration of air near the ceiling is about 5% lower than at the floor. Now look back to Fig. 2.1. The very first fundamental (!) equation of mass transfer shows that there must be a flux of air diffusing from the floor to the ceiling. You may not have realized this before, but every cold floor seems to be producing oxygen...!?

Now the question. The main driving force for mass transfer according to the description used in MASS TRANSFER is the chemical potential gradient (at least in the relatively simple system considered here). Does this theory also predict such oxygen creation? You might first have a look at Fig. 3.6.

Solution

No, the new driving force only contains a mole fraction difference. If the composition of the gas is the same at the floor and the ceiling, there is no driving force for diffusion.

If you want to put this on a firmer basis you may resort to the thermodynamics of irreversible processes. There it is shown that the fundamental driving force for mass transfer is the isothermal chemical potential gradient. For an ideal gas this is equal to:

$$\frac{d\mu_i}{dz} := \frac{d\left[\ln\left[x_i\right]\right]}{dz} \quad \text{or approximately} \quad \frac{Dx_{i1}}{x_i} \cdot \frac{1}{\delta}$$

This is the approximation that we have used in this book. You can see that there are good reasons NOT to use the concentration gradient as the driving force in mass transfer operations.

It should be mentioned that there is another effect involved in the situation analysed here. It is called thermal diffusion. This effect is, however, very small.

4. DIFFUSIVITY OF SPHERES

Calculate the diffusivity of spheres with a diameter of one nanometre in water. Use the model of Figs. 4.1 and 4.2. The viscosity of water is

$$\mu1 := 1 \cdot 10^{-3}$$

Solution

We need the following parameters:

$$R := 8.314 \quad T := 300 \quad L := 6 \cdot 10^{23} \quad d2 := 1 \cdot 10^{-9}$$

This yields:

$$D12 := \frac{R \cdot T}{L \cdot 3 \cdot \pi \cdot \mu1 \cdot d2} \qquad D12 = 4.411 \cdot 10^{-10}$$

This value has the right order of magnitude for the diffusion of a somewhat larger molecule in water.

5. MASS TRANSFER COEFFICIENTS

The figures below give an overview of our knowledge of
coefficients of mass transfer to bubbles, drops and
particles. There are two different cases:
- mobile interfaces (large drops and bubbles) and
- rigid interfaces (solid particles and small drops and
  bubbles).
In either case there are two mass transfer coefficients:
- one for the outside or continuous phase and
- one for the inside or dispersed phase.
For mobile interfaces the flow outside the particle plays a
role in what happens inside. You then see that the physical
properties of the outer phase also enter into the
correlations for the inner phase. So have a good look at
which values you should use.

Estimate the inside and outside mass transfer coefficients
for the following cases:

(a) ion exchange par- ticle in water	diameter density difference dissipation	$d := 10^{-3}$ $d\rho := 100$ $e := 1$
(b,c) a bubble in a clean liquid	diameter (b)	$d := 2 \cdot 10^{-3}$
	diameter (c)	$d := 5 \cdot 10^{-4}$
(d) a drop falling in a gas	diameter (d)	$d := 5 \cdot 10^{-3}$

<u>mass transfer coefficients</u>

<u>rigid interface, outside</u>

small $k = 2\dfrac{D}{d}$

large $k = 0.3\left[\dfrac{g\Delta\rho}{\nu\rho}D^2\right]^{1/3}$

stirred $k = 0.13\dfrac{e^{1/4}D^{2/3}}{\nu^{5/12}}$

dissipation W/kg

take the largest value

<u>rigid interface, inside</u>

$$k = 10\frac{D}{d}$$

inside coefficient

drops and bubbles have
a <u>mobile interface</u> when

$d \geq \left[\dfrac{f\sigma}{g\Delta\rho}\right]^{1/2}$ — interfacial
tension

$f \approx 0.2$ in clean liquids
≈ 1 in dirty liquids
(transition around $d = 1\,mm$)

<u>inside and outside*</u>

$$k = \frac{0.4}{(1+(\rho_d/\rho)^{1/2})^{1/2}}\left[\frac{g^2\Delta\rho^2}{\nu\rho^2}D^3\right]^{1/6}$$

bubbles: 0.40
drops in a liquid: 0.25
drops in a gas: 0.07
*take the rigid value if larger

ρ_d density inside, ρ outside
ν viscosity outside
D ...in the phase considered

You may use the following values for the physical properties:

	gases	liquids	liquid in solid
diffusivity D	10^{-5} □	10^{-9} □	10^{-10} □
density ρ	10^{0} □	10^{3} □	$\sigma := 0.05$ □
dynamic viscosity ν	$2 \cdot 10^{-5}$ □	10^{-6} □	$g := 10$ □

Check whether your results agree with Fig. 4.7. Also calculate the corresponding film thickness and check with Fig. 3.1.

The correlations given are a compilation of an enormous amount of experimental information. Two of the sources are:

P.H.Calderbank and M.B.Moo Young Chem Eng Sci 16 39..54 (1961)

R.Clift, J.R.Grace and M.E.Weber, BUBBLES, DROPS AND PARTICLES, Academic Press, New York, 1978.

Solution

This is simply a matter of using the right data and formulae. The interface is rigid in cases (a) and (c).

For the ion exchange particle (a) I find that the stirred coefficient is the highest: $k := 4.1 \cdot 10^{-5}$
The inside (solid) coefficient is much lower: $k := 1.0 \cdot 10^{-6}$

The bubble (b) is found to have a mobile interface. The outside (liquid) coefficient is: $k := 2.7 \cdot 10^{-4}$
The rigid coefficient has the highest value for the inside: $k := 5.0 \cdot 10^{-2}$

The smaller bubble (c) has a rigid interface. The outside (liquid) coefficient is : $k := 6.5 \cdot 10^{-5}$
The inside coefficient is higher than in case (b): $k := 2.0 \cdot 10^{-1}$

Also the drop has a mobile interface. For the outside (the gas), however, the rigid coefficient has the highest value: $k := 1.1 \cdot 10^{-1}$

The (mobile) inside coefficient is: $k := 2.9 \cdot 10^{-4}$

All these values apply to dilute binary mixtures. In Chapter
8 you will find how you can use the same relations for more
complicated mixtures.

The accuracy of these mass transfer correlations is not
great. You should not be too surprised if they are wrong by a
factor of two. Even larger errors may be expected in the
transition regime between rigid and mobile interfaces.

6. BINARY DIFFUSION WITH A SURFACE REACTION

Carbon monoxide (1) reacts with metallic nickel to form
nickel carbonyl (2), which is a gas.

$$Ni + 4\ CO \longrightarrow Ni(CO)_4$$

The carbonyl diffuses to the bulk of the gas phase. You may
assume the reaction to be infinitely rapid. So there is no CO
at the nickel surface. Parameters given are:

$x1o := 0.50$	$x1\delta := 0.00$	CO mole fractions
$x2o := 0.50$	$x2\delta := 1.00$	carbonyl mole fractions
$k12 := 0.032$		mass transfer coefficient
$c := 37$		total concentration

Calculate the CO flux to the nickel surface.

Solution

This problem is quite straightforward. The mixture is binary,
so we only need one transport relation. We have chosen that
for CO. The bootstrap is given by the reaction stoichiometry,
just as in the example in Fig. 5.5.

auxiliary variables $dx1 := -0.50$ $x1 := 0.25$ $x2 := 0.75$

starting values $v1 := -k12$ $v2 := 0.2 \cdot k12$

given

transport eq. $\dfrac{dx1}{x1} \approx x2 \cdot \dfrac{v2 - v1}{k12}$

bootstrap $v1 \cdot x1 + 4 \cdot v2 \cdot x2 \approx 0$

$\begin{bmatrix} v1 \\ v2 \end{bmatrix} := find(v1, v2)$ $N1 := v1 \cdot c \cdot x1$ $N2 := v2 \cdot c \cdot x2$

$N1 = 0.729$ $N2 = -0.182$

7. CONDENSATION OF A BINARY VAPOUR MIXTURE

A vapour mixture of toluene (1) and di-chloro-ethane (2) has composition $y1o := 0.40$ When it enters the top of a vertical condenser, the liquid that forms has a composition $x1 := 0.30$ The composition of the vapour next to the liquid is given by the equilibrium relation

$$\alpha := 2.14$$

$$y1\delta := \frac{\alpha \cdot x1}{1 + (\alpha - 1) \cdot x1}$$

Other parameters are the mass transfer coefficient and the total concentration:

$$k12 := 0.054 \qquad c := 30$$

(a) Estimate the transfer fluxes of the two components.
(b) Have a look at the fluxes for differing values of x1. There is a value for which both fluxes are zero. What would this imply?

Solution

(a) At the top of the condenser, liquid flows away as it is formed. So the ratio of the fluxes must equal the ratio of the mole fractions in the liquid. This yields the bootstrap relation.

The transport velocities are calculated using the starting values and auxiliary variables

$$v2 := -2 \cdot k12 \qquad\qquad v1 := -k12$$

$$dy1 := y1\delta - y1o \qquad y1 := \frac{y1\delta + y1o}{2} \qquad y2 := 1 - y1$$

given

the transport relation $\qquad \dfrac{dy1}{y1} \approx y2 \cdot \dfrac{v2 - v1}{k12}$

and

the bootstrap relation $\qquad \dfrac{v1 \cdot y1}{v1 \cdot y1 + v2 \cdot y2} \approx x1$

$$\begin{bmatrix} v1 \\ v2 \end{bmatrix} := find(v1, v2) \qquad\qquad \begin{bmatrix} v1 \\ v2 \end{bmatrix} = \begin{vmatrix} 0.021 \\ 0.038 \end{vmatrix}$$

The fluxes then become $\quad \begin{bmatrix} N1 \\ N2 \end{bmatrix} := \begin{bmatrix} v1 \cdot c \cdot y1 \\ v2 \cdot c \cdot y2 \end{bmatrix} \quad \begin{bmatrix} N1 \\ N2 \end{bmatrix} = \begin{vmatrix} 0.274 \\ 0.639 \end{vmatrix}$

(b) Running the above program for
a few values of x1 yields:

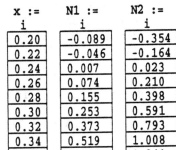

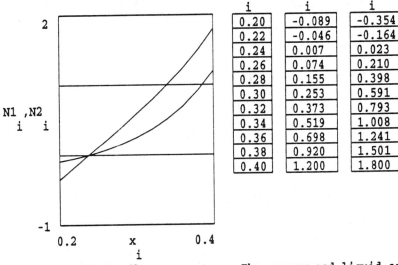

$i := 0 ..10$

$x_i :=$	$N1_i :=$	$N2_i :=$
0.20	-0.089	-0.354
0.22	-0.046	-0.164
0.24	0.007	0.023
0.26	0.074	0.210
0.28	0.155	0.398
0.30	0.253	0.591
0.32	0.373	0.793
0.34	0.519	1.008
0.36	0.698	1.241
0.38	0.920	1.501
0.40	1.200	1.800

At x1 = 0.238 the fluxes are zero. The vapour and liquid are in equilibrium. There is no temperature difference and no heat transfer. Higher values of x1 correspond to condensation, lower ones to evaporation. If you wish to know what happens in a given situation, you must also analyse the heat transfer. More about this in Chapter 7 and in references [2] and [3].

8. THREE GASES

Consider the two bulb diffusion experiment in Fig. 2.3. The system consists of mixtures of hydrogen (1), nitrogen (2) and carbon dioxide (3) at:

$$p := 101 \cdot 10^3 \qquad T := 273 + 35$$

The compositions in the two bulbs A and B are:

$$x1A := 0.00 \quad x2A := 0.50 \quad x3A := 0.50$$

$$x1B := 0.50 \quad x2B := 0.50 \quad x3B := 0.00$$

The diffusivities are:

$$D12 := 83.8 \cdot 10^{-6} \qquad D13 := 68 \cdot 10^{-6} \qquad D23 := 16.8 \cdot 10^{-6}$$

The length of the capillary is $\delta := 86 \cdot 10^{-3}$

Determine the fluxes of the three species assuming quasi-steady state. (This situation will apply if the volumes of the two spheres are large.)

Solution

The mass transfer coefficients are:

$$k12 := \frac{D12}{\delta} \qquad k13 := \frac{D13}{\delta} \qquad k23 := \frac{D23}{\delta}$$

and the total concentration:

$$c := \frac{p}{8.314 \cdot T}$$

Auxiliary variables:

$$x1 := 0.25 \quad x2 := 0.50 \quad x3 := 0.25$$

$$dx1 := -0.50 \qquad dx2 := 0.00$$

Starting values:

$$v1 := -k13 \quad v2 := 0 \quad v3 := k13$$

Given

$$\frac{dx1}{x1} \approx x2 \cdot \frac{v2 - v1}{k12} + x3 \cdot \frac{v3 - v1}{k13}$$

$$\frac{dx2}{x2} \approx x1 \cdot \frac{v1 - v2}{k12} + x3 \cdot \frac{v3 - v2}{k23}$$

$$v1 \cdot x1 + v2 \cdot x2 + v3 \cdot x3 \approx 0$$

$$\begin{bmatrix} v1 \\ v2 \\ v3 \end{bmatrix} := \text{Find}(v1, v2, v3)$$

N1 := v1·x1·c	N1 = 0.0177
N2 := v2·x2·c	N2 = -0.0088
N3 := v3·x3·c	N3 = -0.0088

As you can see, the nitrogen is forced in the same direction as the carbon dioxide, even though its driving force is zero.

9. EQUIMOLAR DISTILLATION OF A TERNARY MIXTURE

Consider distillation of the ternary mixture ethanol (1) -
t-butanol (2) - water (3) in a sieve tray column. The
composition of the vapour leaving the tray (in mole
fractions) is:

$$y1o := 0.5558 \quad y2o := 0.1353 \quad y3o := 1 - y1o - y2o$$

The composition of the vapour in equilibrium with the liquid
on the tray is:

$$y1\delta := 0.6040 \quad y2\delta := 0.1353 \quad y3\delta := 1 - y1\delta - y2\delta$$

The vapour phase mass transfer coefficients are:

$$k12 := 0.08 \quad k23 := 0.16 \quad k13 := 0.21$$

Calculate the species velocities of the three components and
comment on their directions of transfer.

Solution

Auxiliary variables are:

$$dy1 := y1\delta - y1o \qquad y1 := 0.5 \cdot (y1o + y1\delta)$$
$$dy2 := y2\delta - y2o \qquad y2 := 0.5 \cdot (y2o + y2\delta)$$
$$y3 := 1 - y1 - y2$$

Starting values:

$$v1 := -k12 \qquad v2 := 0 \qquad v3 := k12$$

The two transport relations and the relation for equimolar
exchange read:

Given

$$\frac{dy1}{y1} \approx y2 \cdot \frac{v2 - v1}{k12} + y3 \cdot \frac{v3 - v1}{k13}$$

$$\frac{dy2}{y2} \approx y1 \cdot \frac{v1 - v2}{k12} + y3 \cdot \frac{v3 - v2}{k23}$$

$$v1 \cdot y1 + v2 \cdot y2 + v3 \cdot y3 \approx 0$$

$$\begin{bmatrix} v1 \\ v2 \\ v3 \end{bmatrix} := Find(v1, v2, v3) \qquad \begin{bmatrix} v1 \\ v2 \\ v3 \end{bmatrix} = \begin{bmatrix} -0.015 \\ -0.006 \\ 0.034 \end{bmatrix}$$

As you can see, butanol (2) is transferred against its
gradient.

10. NON-EQUIMOLAR DISTILLATION OF A TERNARY MIXTURE

We consider mass transfer during distillation of a ternary
mixture of iso-pentane (1) - ethanol (2) - water (3) in a
sieve tray column. At a certain position in the column, the
bulk vapour composition at a point on a tray is

$\quad$ y1o := 0.62 $\quad$ y2o := 0.16 $\quad$ y3o := 0.22

The corresponding interface mole fractions are

$\quad$ y1δ := 0.60 $\quad$ y2δ := 0.10 $\quad$ y3δ := 0.30

The column operates at a pressure and temperature of

$\quad$ p := 10^5 $\quad$ T := 346

Determine the interfacial molar fluxes of the three
components taking proper account of the differences in the
enthalpies of vaporization of the three species. Examine also
the accuracy of the assumption of equimolar distillation for
this system.

The vapour phase transfer coefficients of the binary pairs
are

$\quad$ k12 := 0.07 $\quad$ k23 := 0.21 $\quad$ k13 := 0.14

The partial molar enthalpies at the prevailing conditions in
the column are:

$\quad$ liquid phase:

$\quad$ H11 := $15.5 \cdot 10^3$ $\quad$ H21 := $10.1 \cdot 10^3$ $\quad$ H31 := $5.0 \cdot 10^3$

$\quad$ vapour phase:

$\quad$ H1v := $38.0 \cdot 10^3$ $\quad$ H2v := $50.6 \cdot 10^3$ $\quad$ H3v := $47.0 \cdot 10^3$

Solution

Auxiliary variables required in the solve block below are:

$\quad$ dy1 := y1δ - y1o $\qquad$ y1 := $0.5 \cdot$ (y1o + y1δ)
$\quad$ dy2 := y2δ - y2o $\qquad$ y2 := $0.5 \cdot$ (y2o + y2δ)
$\qquad\qquad\qquad\qquad\qquad$ y3 := 1 - y1 - y2

Starting values are:

$\quad$ v1 := 0 $\quad$ v2 := k12 $\quad$ v3 := -k13

Given

$$\frac{dy1}{y1} \approx y2 \cdot \frac{v2 - v1}{k12} + y3 \cdot \frac{v3 - v1}{k13}$$

$$\frac{dy2}{y2} \approx y1 \cdot \frac{v1 - v2}{k12} + y3 \cdot \frac{v3 - v2}{k23}$$

The bootstrap is the enthalpy balance:

$$v1 \cdot y1 \cdot (H1v - H1l) + v2 \cdot y2 \cdot (H2v - H2l) \ldots \approx 0$$

$$+ v3 \cdot y3 \cdot (H3v - H3l)$$

$$\begin{bmatrix} v1 \\ v2 \\ v3 \end{bmatrix} := Find(v1, v2, v3) \qquad \begin{bmatrix} v1 \\ v2 \\ v3 \end{bmatrix} = \begin{vmatrix} 0.014 \\ 0.053 \\ -0.043 \end{vmatrix}$$

To obtain the fluxes we need: $\qquad c := \dfrac{p}{8.314 \cdot T}$

N1 := v1·c·y1	N1 = 0.294	
N2 := v2·c·y2	N2 = 0.24	
N3 := v3·c·y3	N3 = -0.389	N1 + N2 + N3 = 0.145

The sum of the three fluxes differs considerably from zero. This is to be expected because of the large differences in the heats of evaporation of the components.

11. DIFFUSION THROUGH TWO STAGNANT GASES

Methane (1) is absorbed from a mixture of argon (2) and
helium (3) by a liquid solvent in a packed column. The liquid
may be assumed to be non-volatile. The column operates at

$$p := 101 \cdot 10^{3} \qquad T := 273 + 25$$

The composition of the bulk vapour at a location in the
column is:

$$y1\delta := 0.4 \qquad y2\delta := 0.4 \qquad y3\delta := 0.2$$

At this location the partial pressure of component (1) at the
interface is

$$p1 := 10.1 \cdot 10^{3}$$

Both argon and helium may be considered as insoluble in the
liquid. The transfer of methane is dictated by the vapour
phase transfer through a film thickness of

$$\delta := 10^{-3}$$

The diffusivities are:

$$D12 := 20.2 \cdot 10^{-6} \qquad D13 := 67.5 \cdot 10^{-6} \qquad D23 := 72.9 \cdot 10^{-6}$$

Determine the transfer flux of methane.

Solution

The mass transfer coefficients are:

$$k12 := \frac{D12}{\delta} \qquad k13 := \frac{D13}{\delta} \qquad k23 := \frac{D23}{\delta}$$

The interface mole fraction of methane is

$$y1o := \frac{p1}{p} \qquad y1o = 0.1$$

The interface compositions of the other two components are
unknown, but their fluxes are zero. We can calculate

$$Dy1 := y1\delta - y1o \qquad y1 := (y1\delta + y1o) \cdot 0.5$$

In the solve block below, there are two unknowns. Starting
values are:

$$y2o := 0.6 \qquad v1 := -k12$$

Given

$$y2 := 0.5 \cdot (y2o + y2\delta)$$
$$y3 := 1 - y1 - y2$$

$$\frac{Dy1}{y1} \approx -y2 \cdot \frac{v1}{k12} + -y3 \cdot \frac{v1}{k13}$$
transport
equation of
component (1)

$$\frac{y2\delta - y2o}{y2} \approx y1 \cdot \frac{v1}{k12}$$
transport
equation of
component (2)

$$\begin{bmatrix} v1 \\ y2o \end{bmatrix} := Find(v1, y2o)$$
$$\begin{bmatrix} v1 \\ y2o \end{bmatrix} = \begin{bmatrix} -0.041 \\ 0.668 \end{bmatrix}$$

and the flux of methane is given by

$$c := \frac{p}{8.314 \cdot T}$$
$$N1 := v1 \cdot c \cdot y1$$
$$N1 = -0.413$$

It is interesting to note that the friction between (2) and (3) is irrelevant here. This is because there is no relative motion between (2) and (3).

12. DEHYDROGENATION OF ETHANOL

This example has been adapted from a wellknown textbook by
Froment and Bischoff (CHEMICAL REACTOR ANALYSIS AND DESIGN,
Wiley 1979, p 151). It is designed to show how to use the M-S
equations for the description of mass transfer with
heterogeneous chemical reactions.

We consider the dehydrogenation of ethanol:

ethanol (1) ----> acetaldehyde (2) + hydrogen (3)

Conditions are:

$$p := 10^5 \qquad T := 548$$

The reaction takes place at a catalyst surface. The reaction
rate is first order in the concentration of ethanol at the
reaction surface:

$$kr := 0.45 \qquad r := -kr \cdot c \cdot x1\delta$$

(this in mol/m2/s). Assume that the intraparticle diffusion
resistance is negligible. Estimate the overall rate of the
reaction of ethanol for the following parameters:

bulk gas phase composition
$x1o := 0.7 \quad x2o := 0.15 \quad x3o := 0.15$

vapour phase transfer coefficients
$k12 := 0.07 \quad k23 := 0.23 \quad k13 := 0.23$

Hint: rewrite the equations in terms of N1.

Solution

This problem is not so simple. We do not know the composition
at the solid surface beforehand. So there are six unknowns:
- the three mole fractions at the surface and
- the three fluxes.
The equations available are:
- two transport relations,
- two stoichiometric ratios,
- one reaction rate equation and
- the summation relation of the three mole
  fractions.
Although they are nothing special, experience shows that it
is difficult to obtain convergence with such a set of
equations using MathCAD. We have therefore rearranged the
equations such as to reduce their number. This has been done
by eliminating the fluxes of (2) and (3).

We take as starting values

$$x1\delta := 0.1$$
$$x2\delta := 0.6 \qquad\qquad c := \dfrac{p}{8.314 \cdot T}$$
$$x3\delta := 0.3$$

for the flux of (1) we estimate

$$N1 := kr \cdot c \cdot x1\delta$$

The rearranged equations yield:

Given

$$x1 := 0.5 \cdot (x1o + x1\delta) \qquad x2 := 0.5 \cdot (x2o + x2\delta)$$
$$x3 := 1 - x1 - x2$$

$$x1\delta - x1o \approx -\left[\frac{x1 + x2}{c \cdot k12} + \frac{x1 + x3}{c \cdot k13}\right] \cdot N1$$

$$x2\delta - x2o \approx \left[\frac{x1 + x2}{c \cdot k12} - \frac{x2 - x3}{c \cdot k23}\right] \cdot N1$$

$$N1 \approx kr \cdot c \cdot x1\delta$$

$$\begin{bmatrix} x1\delta \\ x2\delta \\ N1 \end{bmatrix} := \text{Find}(x1\delta, x2\delta, N1) \qquad\qquad \begin{bmatrix} x1\delta \\ x2\delta \\ N1 \end{bmatrix} = \begin{vmatrix} 0.097 \\ 0.605 \\ 0.959 \end{vmatrix}$$

The "exact" solution of the differential equations yields:

$$N1 = 0.996$$

13. CONDENSATION OF METHANOL

This problem is given in Figs. 7.6 and 7.7. It concerns the
condensation of methanol against a condenser tube in the
presence of nitrogen. There are several transport steps in
this problem:
- heat and mass (methanol) are transported through the gas
  film,
- heat is further transported through the liquid film, the
  tube wall and the coolant.

Only the first step is analysed here. A separate relation is
given for the heat transfer in the second step.

Data: $k12 := 10^{-2}$ mass transfer coefficient, m/s

        $h := 12$ heat transfer coefficient, W/(m2 K)

        $To := 345$ bulk gas temperature, K

        $y1o := 0.40$ bulk mole fraction nitrogen

        $y2o := 0.60$ bulk mole fraction methanol

        $H1(T) := 29 \cdot T$ enthalpies, J/mol

        $H2(T) := 45 \cdot T + 36000$

$$y2\delta(T) := \exp\left[4380 \cdot \left[\frac{1}{338} - \frac{1}{T}\right]\right] \quad \begin{array}{l}\text{mole fraction methanol} \\ \text{at interface}\end{array}$$

        $E(N2,T\delta) := N2 \cdot H2(T\delta) + 400 \cdot (T\delta - 290)$
                             relation between $T\delta$, N2 and the energy
                             flux to the coolant.

Find the mass and heat flux and the interfacial temperature
and composition. Before you start MathCAD, first write down
which equations you need to solve! It may be handy to write
the transport relation in terms of fluxes.

Solution

You have to find the value of $T\delta$ and N2 simultaneously. The
equations governing the problem are listed below. The last
equation is just a rearrangement of the transport equation in
terms of the flux N2.

Initialize: $T\delta$:= 340   N2 := 0

given
$$T := \frac{To + T\delta}{2}$$

$$y1\delta := 1 - y2\delta(T\delta)$$

$$y2 := (y2o + y2\delta(T\delta)) \cdot 0.5$$

$$y1 := (y1o + y2o) \cdot 0.5$$

$$Dy2 := (y2\delta(T\delta) - y2o)$$

$$h \cdot (To - T\delta) + N2 \cdot H2(T) \approx N2 \cdot H2(T\delta) + 400 \cdot (T\delta - 290)$$

$$Dy2 \approx -y1 \cdot \frac{N2}{k12}$$

$$\begin{bmatrix} T\delta \\ N2 \end{bmatrix} := find(T\delta, N2)$$

$$\begin{bmatrix} T\delta \\ N2 \end{bmatrix} = \begin{vmatrix} 291.629 \\ 0.009 \end{vmatrix}$$

The energy flux follows from:

$$E := N2 \cdot H2(T\delta) + 400 \cdot (T\delta - 290)$$
$$E = 1.116 \cdot 10^{3}$$

Note that the value does depend on the zero point chosen for the enthalpy.

14. COMPOSITION PROFILES IN A NON-IDEAL BINARY MIXTURE

We consider the simple non-ideal binary mixture given in Figs. 8.1 and 8.6. The two components are diffusing through a thin film: the fluxes are equal but opposite. The liquid on the left has a composition $x1o$ = 1.00; for that on the right, $x1\delta$ = 0.00. You may assume the MS diffusivity and total concentration to be independent of composition.

Calculate the composition profiles for different values of the non-ideality parameter A:

$$i := 0 .. 2 \qquad A :=$$

		A_i
ideal..................		0
non-ideal...............		1
with a consolute point...		2

Hint: this is most easily approached via the Fick equation in Fig. 8.6.

What happens at the consolute point? Explain.

Solution

As $N1 := - N2$, the total flux is zero. So $N1 = J1$ and the Fick equation can be written as

$$N1 := -Dms \cdot c \cdot (1 - 2 \cdot A \cdot x1 \cdot (1 - x1)) \cdot \frac{dx1}{dz} \quad \Box$$

The flux is a constant, and integration yields:

$$x1 - A \cdot x1^2 + \frac{2}{3} \cdot A \cdot x1^3 + const := \frac{-(N1 \cdot \delta)}{c \cdot D12} \cdot z' \quad \Box \text{ with } z' := \frac{z}{\delta} \quad \Box$$

The constant and the flux follow from the boundary conditions.
The profile becomes

$$z'(x,A) := 1 + \left[x - A \cdot x^2 + \frac{2 \cdot A}{3} \cdot x^3 \right] \cdot \left[\frac{A}{3} - 1 \right]^{-1}$$

which is plotted below.

$$j := 0 .. 100 \qquad x_j := 0.01 \cdot j$$

$$z'0_j := z'\left[x_j, 0 \right] \qquad z'1_j := z'\left[x_j, 1 \right] \qquad z'2_j := z'\left[x_j, 2 \right]$$

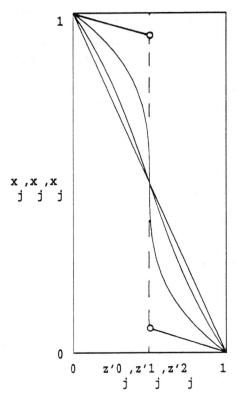

x_j, x_j, x_j

$z'0_j, z'1_j, z'2_j$

For an ideal solution
(A = 0) we obtain a linear
profile as expected. With
increasing non-ideality
the profiles become more
and more s-shaped.

For A = 2 the profile is
vertical at $x1 = 0.5$ (the
consolute point, where the
Fick diffusivity is zero).
Note that the zero diffu-
sivity does not imply that
all mass transfer ceases!

For A > 2 the solution
forms two phases, with an
interface at $z' = 0.5$.
There are then two sepa-
rate parts to the profile.
These have different
boundary conditions and
constants. I have plotted
the result for A = 3. We
will not analyse this
situation further here.

15. MASS TRANSFER IN A TERNARY NON-IDEAL LIQUID MIXTURE.

This example has been adapted from R. Krishna et al., Chem.Eng. Sci., 40, 893 (1985). It concerns mass transfer in the glycerol-rich phase of the mixture glycerol (1) - water (2) and acetone (3) (see Fig. 8.8). The compositions on the two sides of a film and their average values are:

$$x1o := 0.5480 \qquad x2o := 0.2838$$
$$x1a := 0.6652 \qquad x2a := 0.2358$$
$$x1\delta := 0.7824 \qquad x2\delta := 0.1877$$

The mass transfer coefficients are:

$$k12 := 1.1 \cdot 10^{-4} \qquad k13 := 1.8 \cdot 10^{-4} \qquad k23 := 5.6 \cdot 10^{-4}$$

The activity coefficients at the two sides of the film and at the average composition are:

$$\tau1o := 0.7440 \qquad \tau2o := 0.2519$$
$$\tau1a := 0.8080 \qquad \tau2a := 0.1764$$
$$\tau1\delta := 0.8776 \qquad \tau2\delta := 0.1235$$

The mass transfer resistance in this system is mainly in the (viscous) glycerol-rich phase. This implies that the composition in the acetone-rich phase must closely follow the boundary of the demixing zone. As you can see in Fig. 8.8 this implies that $x1' = 0.04$ (approximately). So

$$\frac{N1}{N1 + N2 + N3} := 0.04 \;\square$$

You may use this as a bootstrap relation. Calculate the transfer velocities of the three species. Does the non-ideality have an important effect?

Solution

The transport equations are given below. The bootstrap is a modified form of the relation given.

$$x3a := 1 - x1a - x2a$$

Initial values: $v1 := 0 \qquad v2 := 0 \qquad v3 := 0$

Given

$$\frac{\tau1\delta \cdot x1\delta - \tau1o \cdot x1o}{\tau1a \cdot x1a} \approx x2a \cdot \frac{v2 - v1}{k12} + x3a \cdot \frac{v3 - v1}{k13}$$

$$\frac{\tau2\delta \cdot x2\delta - \tau2o \cdot x2o}{\tau2a \cdot x2a} \approx x1a \cdot \frac{v1 - v2}{k12} + x3a \cdot \frac{v3 - v2}{k23}$$

$$v1 \cdot x1a \approx 0.04 \cdot (v1 \cdot x1a + v2 \cdot x2a + v3 \cdot x3a)$$

$$\text{Find}(v1, v2, v3) = \begin{vmatrix} 0.000004 \\ 0.000196 \\ 0.000199 \end{vmatrix}$$

Notice that all three velocities are positive.

The non-ideality has a substantial effect, as you can see by comparing the two driving forces:

$$\frac{\tau1\delta \cdot x1\delta - \tau1o \cdot x1o}{\tau1a \cdot x1a} = 0.519 \qquad \frac{x1\delta - x1o}{x1a} = 0.352$$

$$\frac{\tau2\delta \cdot x2\delta - \tau2o \cdot x2o}{\tau2a \cdot x2a} = -1.161 \qquad \frac{x2\delta - x2o}{x2a} = -0.408$$

16. TEST OF THE VIGNES RELATIONSHIP

Tyn and Calus (J.Chem.Eng.Data, 20, 310 (1975)) have measured the Fick diffusivity in the system ethanol (1) - water (2) at 40 C at various compositions. Their data are given below.

mole fraction   Fick diffusivity (*10^9 m2 s-1)

$i := 0 \,..18$ $x1 :=$
i

$D12 :=$
i

$x1_i$	$D12_i$
0.000	1.700
0.024	1.510
0.100	1.000
0.144	0.780
0.200	0.680
0.254	0.635
0.300	0.610
0.400	0.640
0.500	0.730
0.590	0.850
0.600	0.865
0.680	1.020
0.700	1.060
0.792	1.260
0.800	1.275
0.880	1.440
0.900	1.475
0.960	1.570
1.000	1.640

Test the accuracy of the Vignes model using the above data. (This is the logarithmic interpolation formula for the MS diffusivities given in Fig. 9.5). The Van Laar coefficients for this system are given by

A12 := 1.4599   A21 := 0.9609

The thermodynamic correction factor can be calculated from the Van Laar constants as follows:

$$\Gamma_i := 1 + x1_i \cdot \frac{d}{dx1_i} \ln\left[\tau_i\right] \quad \text{or with} \quad x2_i := 1 - x1_i$$

$$\Gamma_i := 1 - 2 \cdot x1_i \cdot x2_i \cdot \frac{(A12 \cdot A21)^2}{\left[A21 \cdot x2_i + A12 \cdot x1_i\right]^3}$$

Solution

The **MS** coefficients follow from:

$$Dms_i := \frac{D12_i}{\Gamma_i}$$

They are plotted below. They indeed follow the Vignes relation fairly well.

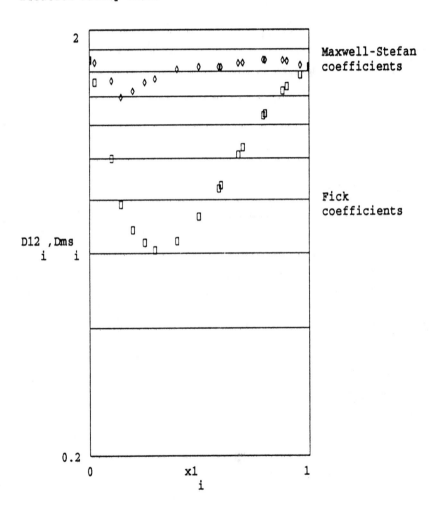

17. DIFFUSION NEAR THE CONSOLUTE POINT

Clark and Rowley (AIChE J., 32, 1125 (1986)) have published
experimental data on the Fick diffusivity near the region of
the liquid-liquid critical point of methanol (1) - n-hexane
(2) mixtures. A portion of their experimental data for the
temperature of 40 C is given below.

i := 0 ..14   mole fraction   Fick diffusivity (10^{-10} m2 s-1)

$x1_i :=$ $Df_i :=$

$x1_i$	Df_i
0.0643	30.84
0.1245	19.37
0.2307	10.95
0.4062	3.751
0.4979	2.092
0.5331	1.939
0.5353	2.154
0.5610	1.678
0.5917	2.055
0.6408	2.867
0.7276	4.674
0.7289	4.632
0.8324	9.986
0.9135	18.26
0.9903	30.85

Investigate the composition dependance of the MS diffusivity
and test the accuracy of the Vignes relationship. The
non-ideality of this system can be described by the NRTL
equations with the parameters:

$$T := 273 + 40 \qquad R := 8.314 \qquad \alpha := 0.2$$

$$\tau12 := \frac{1540.70}{T} - 3.7096 \qquad \tau21 := \frac{1606.53}{T} - 4.2156$$

$$G12 := \exp(-\alpha \cdot \tau12) \qquad G21 := \exp(-\alpha \cdot \tau21)$$

$$x2_i := 1 - x1_i$$

$$\Gamma_i := 1 - 2 \cdot x1_i \cdot x2_i \cdot \left[\tau21 \cdot \frac{G21^2}{\left[x1_i + x2_i \cdot G21\right]^3} \cdots + \tau12 \cdot \frac{G12^2}{\left[x2_i + x1_i \cdot G12\right]^3} \right]$$

Solution

The MS diffusivities are given by:

$$Dms_i := \frac{Df_i}{\Gamma_i}$$

They are plotted below. They do vary somewhat, but much less than the Fick coefficients. They follow the Vignes relation approximately.

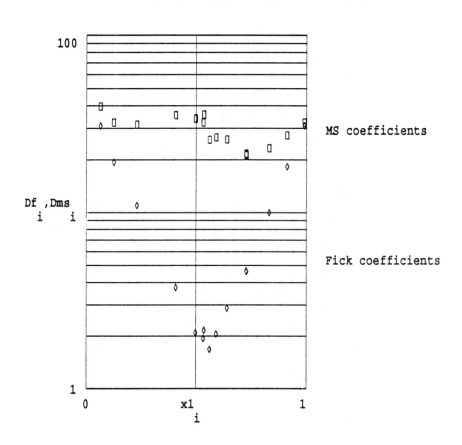

MS coefficients

Fick coefficients

18. MULTICOMPONENT VIGNES

Determine the MS diffusivities for the system acetone (1) - benzene (2) - carbon tetrachloride (3) at 25 C. Estimates of the infinite dilution coefficients are given below.

$$D12_1 := 4.15 \quad D12_2 := 2.75$$
$$D13_1 := 3.57 \quad D13_3 := 1.70 \qquad \text{units} \quad 10^{-9} \cdot m^2 \cdot s^{-1} \; \square$$
$$D23_2 := 1.91 \quad D23_3 := 1.42$$

The subscripts have the following meaning:

$Dij_1 \; \square$ means at $x1 := 1$

$Dij_2 \; \square$ at $x2 := 1$ and

$Dij_3 \; \square$ at $x3 := 1$.

The liquid composition is:

$$x1 := 0.7 \quad x2 := 0.15 \quad x3 := 0.15$$

Solution

We use the ternary Vignes model as given in Fig. 9.6.

$$D12 := D12_1^{\,x1+0.5\cdot x3} \cdot D12_2^{\,x2+0.5\cdot x3} \qquad D12 = 3.783$$

$$D23 := D23_2^{\,x2+0.5\cdot x1} \cdot D23_3^{\,x3+0.5\cdot x1} \qquad D23 = 1.647$$

$$D13 := D13_1^{\,x1+0.5\cdot x2} \cdot D13_3^{\,x3+0.5\cdot x2} \qquad D13 = 3.021$$

19. DIFFUSIVITY OF A TRACE IN A LIQUID MIXTURE

Holmes, Olander and Wilke (AIChE J., 8, 646 (1962)) have
reported data on the diffusion coefficient of trace amounts
of toluene (1) in a mixture of n-tetradecane (2) and n-hexane
(3) at a temperature of 25 C. They measured the toluene
diffusivity in varying compositions of the binary mixture of
(2) and (3) and their data is reproduced below.

mole fraction   effective diffusivity

$i := 1 .. 8$ $x2_i :=$ $D1_i :=$

$x2_i$
1.000
0.803
0.672
0.501
0.336
0.215
0.113
0.000

$D1_i$
1.08
1.37
1.58
1.92
2.38
2.90
3.57
4.62

$10^{-9} \cdot \dfrac{m^2}{s}$ □

The effective diffusivity is defined by:

$$N1 := -c \cdot D1 \cdot \frac{dx1}{dz} \quad □$$

Set up the MS equations to describe the diffusion of toluene
in the binary mixture of n-tetradecane and n-hexane and
rationalize the experimental results. You may regard the
mixture as ideal.

Solution

The MS equation for toluene reads:

$$\frac{1}{x1} \cdot \frac{dx1}{dz} := x2 \cdot \frac{-v1}{D12} + x3 \cdot \frac{-v1}{D13} \quad □ \quad \text{or} \quad N1 := -c \cdot \left[\frac{x2}{D12} + \frac{x3}{D13} \right]^{-1} \cdot \frac{dx1}{dz} \quad □$$

So
$$D1 := \left[\frac{x2}{D12} + \frac{x3}{D13} \right]^{-1} \quad □$$

This gives the values at the
extreme compositions:

for $x3 := 0 \;$ ---> $\; D12 := 1.08$
for $x2 := 0 \;$ ---> $\; D13 := 4.68$

For lack of something better, we assume that these values are
independent of composition. The calculated and measured
effective diffusivities are plotted below.

$$j := 0 \ ..100$$

$$x2'_j := \frac{j}{100}$$

$$D1'_j := \left[\frac{x2'_j}{D12} + \frac{1 - x2'_j}{D13} \right]^{-1}$$

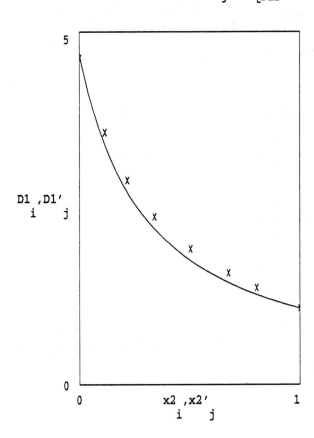

20. ULTRACENTRIFUGATION OF AN AQUEOUS SOLUTION OF ALBUMIN

A dilute aqueous solution of albumin is placed in a short
centrifuge tube and rotated until a steady state is obtained.

Data $M1 := 45$ kg mol-1 $R := 8.314$
        $\rho1 := 1340$ kg m-3 $T := 297$

        $V1 := \dfrac{M1}{\rho1}$ m3 mol-1 $ro := 0.099$

        $\rho := 1000$ kg m-3 $r\delta := 0.100$

        $\omega := 2 \cdot \pi \cdot \left[\dfrac{10000}{60}\right]$ s-1

Estimate the thickness of the protein-containing layer at the
rim of the centrifuge.

Solution

This is the case of the right side of Fig. 10.5. The
thickness is:

$$\delta r := \frac{2 \cdot R \cdot T}{V1 \cdot (\rho1 - \rho) \cdot \omega^2 \cdot r\delta} \qquad \delta r = 0.004$$

The complete concentration profile is of course more or less
exponential.

21. ULTRACENTRIFUGATION OF A BINARY MIXTURE

An equimolar mixture of benzene (1) and carbon tetrachloride
(2) is placed in a sedimentation cell in an ultracenrifuge
and rotated at 30 000 rpm. The outer radius of the cell is
100 mm and the depth of the liquid in the cell is 40 mm. The
temperature is 40 C.

Data: $V1 := 89 \cdot 10^{-6}$ m3 mol-1 $R := 8.314$ $ro := 0.06$

$\rho := 1252$ kg m-3 $T := 273 + 40$ $r\delta := 0.10$

$M1 := 0.07811$ kg mol-1   30000 $ra := 0.08$

$M2 := 0.15384$ kg mol-1 $\omega := 2 \cdot \pi \cdot \dfrac{30000}{60}$

$\rho1 := \dfrac{M1}{V1}$ kg m-3 $\rho1 = 877.64$

The activity coefficient of benzene in the solution is given
by

$$\tau1(x1) := \exp\left[0.14 \cdot (1 - x1)^2\right]$$

Estimate the composition variation of (1) over the centrifuge
at steady state.

Solution

We are clearly dealing with the situation on the left-hand
side of Fig. 10.5.

The centrifugal-plus-pressure term is

$$r\%1 := \frac{V1 \cdot (\rho1 - \rho) \cdot \omega^2 \cdot ra \cdot (r\delta - ro)}{R \cdot T}   \qquad r\%1 = -0.404$$

The activity term has to include non-ideality. The mole
fraction difference over the centrifuge is then obtained by
solving the following set of equations

Average (given): $x1 := 0.50$

Initial values: $xlo := 0.52$ $x1\delta := 0.48$

Given

$$\frac{\tau1(x1\delta) \cdot x1\delta - \tau1(xlo) \cdot xlo}{\tau1(x1) \cdot x1} \approx r\%1$$

$$xlo + x1\delta \approx 1$$

$$\text{Find}(xlo, x1\delta) = \begin{vmatrix} 0.609 \\ 0.391 \end{vmatrix}$$

22. DIFFUSION IN MIXED ELECTROLYTES

Consider a cell with two stirred compartments separated by a diaphragm. The left compartment is labelled A and the right one B. The diaphragm is an open porous plate, with an effective thickness $\delta := 0.001$

The volume of the diaphragm is negligible compared to that of the compartments. You may therefore regard the transport as steady. Compartment A contains a dilute mixture of HCl and NaCl. Compartment B contains pure water. The ions H+, Na+ and Cl- are labelled (1), (2) and (3) respectively. The total concentration is the same in a series of experiments:

$$x1 + x2 := 10^{-4}$$

The infinite dilution diffusivities of H+, Na+ and Cl- are:

$$D1w := 9.3 \cdot 10^{-9} \qquad D2w := 1.33 \cdot 10^{-9} \qquad D3w := 2.03 \cdot 10^{-9}$$

This yields the mass transfer coefficients:

$$k1w := \frac{D1w}{\delta} \qquad k2w := \frac{D2w}{\delta} \qquad k3w := \frac{D3w}{\delta}$$

Investigate the transfer velocities of the ions and the potential difference across the diaphragm for a range of ratios of the H+ and Na+ concentrations:

$$10^{-3} < r < 10^{3}$$

Solution

As an example we take $r := 1$. The mole fractions are then:

$$x1 := 0.5 \cdot 10^{-4} \qquad x2 := 0.5 \cdot 10^{-4} \qquad x3 := 10^{-4}$$

The species velocities and potential difference follow from the equations below with the initial values:

$$v1 := 0 \qquad v2 := 0 \qquad v3 := 0 \qquad D\phi := 0$$

Given

$$-2 + 40 \cdot D\phi \approx \frac{-v1}{k1w} \qquad -2 + 40 \cdot D\phi \approx \frac{-v2}{k2w} \qquad -2 - 40 \cdot D\phi \approx \frac{-v3}{k3w}$$

$$v1 \cdot x1 + v2 \cdot x2 \approx v3 \cdot x3 \qquad \text{no currrent}$$

$$\begin{bmatrix} v1 \\ v2 \\ v3 \\ D\phi \end{bmatrix} := \text{Find}(v1,v2,v3,D\phi) \qquad \begin{bmatrix} v1 \\ v2 \\ v3 \end{bmatrix} \cdot 10^{6} = \begin{vmatrix} 10.281 \\ 1.47 \\ 5.876 \end{vmatrix}$$

$$D\phi = 0.022$$

$k := 0 ..9$

$r :=$ k	$v1 :=$ k	$v2 :=$ k	$v3 :=$ k	$D\phi :=$ k
0.001	22.4	3.21	3.23	-0.010
0.01	22.0	3.14	3.33	-0.009
0.03	21.0	3.01	3.53	-0.007
0.1	18.5	2.64	4.08	0.000
0.3	14.5	2.08	4.95	0.011
1	10.3	1.47	5.88	0.022
3	8.09	1.16	6.36	0.028
10	7.12	1.02	6.57	0.031
100	6.71	0.96	6.66	0.032
1000	6.67	0.95	6.67	0.032

$y := 0$
$\quad k$

The results of a number of calculations are plotted below.

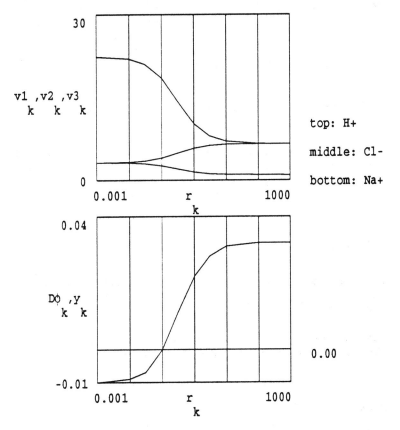

top: H+

middle: Cl-

bottom: Na+

For low H+ concentrations (small r) the electrical potential difference is caused by a faster movement of Cl- compared to Na+. This potential difference increases the speed of the trace of H+. For high H+ concentrations the movement of the H+ causes a positive potential difference which slows H+ down.

Note that for a small r the Na+ and Cl- move with the same velocity. For a large r the H+ and Cl- move along together.

23. CONDUCTION IN AN ELECTROLYTE

Calculate the transport velocities of H+ and Cl- in the
conductivity cell of Fig. 11.7. Vary the mole fraction salt
between 0.0002 and 0.02. You may use the diffusivities given
in Figs. 11.8 and 11.9.

Solution

There are three components: H+ (1), Cl- (2) and H2O (3). The
mole fractions are:

$$x1 := 0.01 \quad x2 := x1 \quad x3 := 1 - x1 - x2$$

For the diffusivities we find in Figs. 10.8 and 10.9:

$$D13 := 9.3 \cdot 10^{-9} \qquad D23 := 2.0 \cdot 10^{-9}$$

$$D12 := 5 \cdot 10^{-9} \cdot \sqrt{x1} \quad D12 = 5 \cdot 10^{-10}$$

(The formula in Fig. 11.8 is not very accurate for H+
containing systems).

With a film thickness $\delta := 10^{-2}$ this yields the mass
transfer coefficients:

$$k12 := \frac{D12}{\delta} \quad k13 := \frac{D13}{\delta} \quad k23 := \frac{D23}{\delta}$$

The potential difference is: $D\phi := 1$

and the transport relations with their initial values read:

$$v1 := 0 \quad v2 := 0$$

given

$$40 \cdot D\phi \approx x2 \cdot \frac{v2 - v1}{k12} + x3 \cdot \frac{-v1}{k13} \qquad -40 \cdot D\phi \approx x1 \cdot \frac{v1 - v2}{k12} + x3 \cdot \frac{-v2}{k23}$$

$$\begin{bmatrix} v1 \\ v2 \end{bmatrix} := \text{find}(v1, v2) \qquad \begin{bmatrix} v1 \\ v2 \end{bmatrix} = \begin{vmatrix} -0.00003085 \\ 0.00000663 \end{vmatrix}$$

This has been calculated
for a number of cases:

The results are plotted
below in the manner of
Fig. 11.7.

k := 1 ..6

$x1_k :=$	$v1_k :=$	$v2_k :=$
0.0001	-36.4	7.83
0.0003	-35.8	7.70
0.001	-34.8	7.48
0.003	-33.3	7.16
0.01	-30.8	6.63
0.03	-27.9	6.00

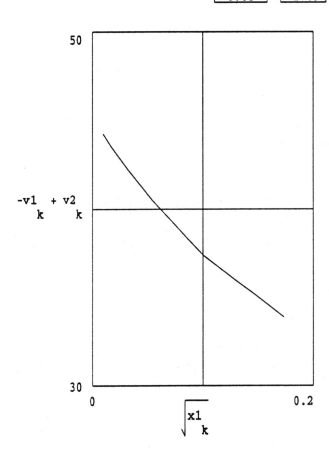

$-v1_k + v2_k$

$\sqrt{\dfrac{x1}{k}}$

24. A TERNARY GAS MIXTURE THROUGH A POROUS MEDIUM

Consider the diffusion of a ternary gas mixture He (1) - Ne (2) - Ar (3) through an inert porous membrane. We model the membrane as being made of parallel cylindrical pores. We will have a look at three different pore diameters: 100, 200 and 300 nm. The other parameters are:

upstream pressure: $po := 210 \cdot 10^3$

downstream pressure: $p\delta := 190 \cdot 10^3$

temperature: $T := 300$ $R := 8.314$

molar masses: $M1 := 0.004$
                             $M2 := 0.020183$
                             $M3 := 0.039948$

gas viscosity: $\mu := 22 \cdot 10^{-6}$

membrane thickness: $\delta := 9.6 \cdot 10^{-3}$

the compositions on the two sides of the membrane are:

$x1o := 0.4$ $x1\delta := 0.4$
$x2o := 0.4$ $x2\delta := 0.3$
$x3o := 0.2$ $x3\delta := 0.3$

the bulk gas diffusivities at $pr := 100 \cdot 10^3$ are:

$D12 := 1.068 \cdot 10^{-4}$ $D13 := 0.724 \cdot 10^{-4}$ $D23 := 0.316 \cdot 10^{-4}$

Determine the magnitudes and directions of the velocities of the three species for the different diameters. Try to explain your results.

Solution

The required equations can be found in Figs. 13.1 and 13.2.

As an example we consider the case with a pore diameter $do := 100 \cdot 10^{-9}$

First we calculate the average and partial pressures:

$$p := \frac{po + p\delta}{2} \qquad p1 := \frac{po \cdot x1o + p\delta \cdot x1\delta}{2}$$

$$p2 := \frac{po \cdot x2o + p\delta \cdot x2\delta}{2} \qquad p3 := \frac{po \cdot x3o + p\delta \cdot x3\delta}{2}$$

Then the pressure differences:

$dp := p\delta - po$

```
dp1 := pδ·x1δ - po·x1o
dp2 := pδ·x2δ - po·x2o
dp3 := pδ·x3δ - po·x3o
```

and the average mole fractions:

```
x1 := 0.5·(x1o + x1δ)
x2 := 0.5·(x2o + x2δ)
x3 := 0.5·(x3o + x3δ)
```

The gas/membrane diffusivities follow from:

$$D1 := \sqrt{\frac{8\cdot R\cdot T}{9\cdot \pi \cdot M1}}\cdot do \qquad D2 := \sqrt{\frac{8\cdot R\cdot T}{9\cdot \pi \cdot M2}}\cdot do \qquad D3 := \sqrt{\frac{8\cdot R\cdot T}{9\cdot \pi \cdot M3}}\cdot do$$

and the mass transfer coefficients become:

$$k12 := \frac{D12}{\delta}\cdot\left[\frac{pr}{p}\right] \qquad k13 := \frac{D13}{\delta}\cdot\left[\frac{pr}{p}\right] \qquad k23 := \frac{D23}{\delta}\cdot\left[\frac{pr}{p}\right]$$

$$k1 := \frac{D1}{\delta} \qquad k2 := \frac{D2}{\delta} \qquad k3 := \frac{D3}{\delta}$$

The viscous flow velocity is:

$$u := \frac{-1}{32}\cdot\frac{dp\cdot do^2}{\mu\cdot\delta}$$

The diffusion velocities then follow from:

initial values   v1 := k1   v2 := k2   v3 := -k3

Given

$$\frac{dp1}{p1} \approx x2\cdot\frac{v2 - v1}{k12} + x3\cdot\frac{v3 - v1}{k13} - \frac{v1}{k1}$$

$$\frac{dp2}{p2} \approx x1\cdot\frac{v1 - v2}{k12} + x3\cdot\frac{v3 - v2}{k23} - \frac{v2}{k2}$$

$$\frac{dp3}{p3} \approx x1\cdot\frac{v1 - v3}{k13} + x2\cdot\frac{v2 - v3}{k23} - \frac{v3}{k3}$$

$$\begin{bmatrix} v1 \\ v2 \\ v3 \end{bmatrix} := Find(v1,v2,v3) \qquad \begin{bmatrix} v1 \\ v2 \\ v3 \end{bmatrix} = \begin{vmatrix} 0.000344 \\ 0.000522 \\ -0.000149 \end{vmatrix}$$

total species
velocities:

$$\begin{bmatrix} w1 \\ w2 \\ w3 \end{bmatrix} := \begin{bmatrix} v1 \\ v2 \\ v3 \end{bmatrix} + u \qquad \begin{bmatrix} w1 \\ w2 \\ w3 \end{bmatrix} = \begin{vmatrix} 0.000373 \\ 0.000552 \\ -0.00012 \end{vmatrix}$$

The results for the three cases have been calculated
separately:
k := 1 ..3

do :=	w1 :=	w2 :=	w3 :=
k	k	k	k
100	373	552	-12
200	730	982	38
300	1123	1414	323

do in nm,

w in $\mu m\ s^{-1}$

The transport direction of Argon depends on the pore
diameter! This is due to two effects:
- in narrow pores the friction of Ar with the membrane
  dominates and Ar diffuses down its own gradient,
- in wider pores the relative increase of the friction
  between the gases and the larger contribution of viscous
  flow cause it to go the other way.

Although helium has no gradient it is transported along with
the neon by gas/gas friction and viscous flow.

25. A NOVEL PROCESS FOR GAS SEPARATIONS

A student having followed a course by Wesselingh and Krishna on the MS approach to mass transfer has come up with an idea for selective removal of styrene (1) from a mixture with ethyl benzene (2) and hydrogen (3). You are required to examine his invention and comment on its feasibility.

The ternary mixture is made to diffuse through an inert porous membrane. The membrane is made up of parallel cylindrical pores. Important parameters are:

pore diameter $\qquad$ $do := 1.5 \cdot 10^{-6}$

upstream pressure $\qquad$ $po := 202.5 \cdot 10^{3}$

downstream pressure $\qquad$ $p\delta := 197.5 \cdot 10^{3}$

temperature $\qquad$ $T := 500 \quad R := 8.314$

compositions:

upstream $\quad x1o := 0.02 \quad x2o := 0.28 \quad x3o := 0.70$

downstream $\quad x1\delta := 0.10 \quad x2\delta := 0.25 \quad x3\delta := 0.65$

membrane thickness $\qquad$ $\delta := 10^{-2}$

molar masses $M1 := 0.104 \quad M2 := 0.106 \quad M3 := 0.002$

gas viscosity $\qquad$ $\mu := 30 \cdot 10^{-6} \; \left(Kg/m \cdot s \right)$

diffusivities in the bulk gas at $pr := 101.3 \cdot 10^{3}$

$$D12 := 7.03 \cdot 10^{-6} \qquad D13 := 77 \cdot 10^{-6} \qquad D23 := 75.9 \cdot 10^{-6}$$

The basis of the invention according to the student is that the flux of styrene under the above set of conditions is directed from the upstream towards the downstream. That is against the intrinsic force. Therefore styrene can be concentrated from a lower to a higher level.

Using the parallel pore model described in Figs. 13.1 and 13.2 you are required to see if indeed uphill transport can be effected this way. Also examine how the system can be developed in practice.

Solution

First of all we will check whether styrene does move in the
predicted direction. For this purpose we need to know a
number of intermediate values:

pressure differences:

$$dp := p\delta - po$$
$$dp1 := p\delta \cdot x1\delta - po \cdot x1o$$
$$dp2 := p\delta \cdot x2\delta - po \cdot x2o$$
$$dp3 := p\delta \cdot x3\delta - po \cdot x3o$$

average pressures:

$$p := 0.5 \cdot (p\delta + po)$$
$$p1 := 0.5 \cdot (p\delta \cdot x1\delta + po \cdot x1o)$$
$$p2 := 0.5 \cdot (p\delta \cdot x2\delta + po \cdot x2o)$$
$$p3 := 0.5 \cdot (p\delta \cdot x3\delta + po \cdot x3o)$$

average mole fractions:

$$x1 := 0.5 \cdot (x1o + x1\delta)$$
$$x2 := 0.5 \cdot (x2o + x2\delta)$$
$$x3 := 0.5 \cdot (x3o + x3\delta)$$

membrane diffusivities:

$$D1 := \sqrt{\frac{8}{9 \cdot \pi} \cdot \frac{R \cdot T}{M1}} \cdot do \qquad D2 := \sqrt{\frac{8}{9 \cdot \pi} \cdot \frac{R \cdot T}{M2}} \cdot do \qquad D3 := \sqrt{\frac{8}{9 \cdot \pi} \cdot \frac{R \cdot T}{M3}} \cdot do$$

mass transfer coefficients:

$$k12 := \frac{D12}{\delta} \cdot \frac{pr}{p} \qquad k13 := \frac{D13}{\delta} \cdot \frac{pr}{p} \qquad k23 := \frac{D23}{\delta} \cdot \frac{pr}{p}$$

$$k1 := \frac{D1}{\delta} \qquad k2 := \frac{D2}{\delta} \qquad k3 := \frac{D3}{\delta}$$

The diffusion velocities follow from the equations in the
following solve block:

$$v1 := 0 \qquad v2 := 0 \qquad v3 := 0 \qquad \text{initial estimates}$$

Given

$$\frac{dp1}{p1} \approx x2 \cdot \frac{v2 - v1}{k12} + x3 \cdot \frac{v3 - v1}{k13} + \frac{-v1}{k1}$$

$$\frac{dp2}{p2} \approx x1 \cdot \frac{v1 - v2}{k12} + x3 \cdot \frac{v3 - v2}{k23} + \frac{-v2}{k2}$$

$$\frac{dp3}{p3} \approx x1 \cdot \frac{v1 - v3}{k13} + x2 \cdot \frac{v2 - v3}{k23} + \frac{-v3}{k3}$$

$$\begin{bmatrix} v1 \\ v2 \\ v3 \end{bmatrix} := \text{Find}(v1, v2, v3) \qquad \begin{bmatrix} v1 \\ v2 \\ v3 \end{bmatrix} = \begin{vmatrix} -0.00034 \\ 0.00093 \\ 0.001698 \end{vmatrix}$$

To these velocities we must add the viscous flow velocity

$$u := \frac{-1}{32} \cdot \frac{dp \cdot do^2}{\mu \cdot \delta} \qquad u = 0.001$$

So the total velocities become:

$$w := \begin{bmatrix} v1 \\ v2 \\ v3 \end{bmatrix} + u \qquad w = \begin{vmatrix} 0.000832 \\ 0.002102 \\ 0.00287 \end{vmatrix}$$

The styrene is transferring against its gradient. The styrene transport is, however, mainly by viscous flow. A simple calculation shows that you are then diluting and not concentrating the styrene. So there is no invention.

26. DIALYSIS OF HYDROCARBONS

Dialysis experiments with hydrocarbons are reported by J.G.A.
Bitter in: Transport mechanisms in membrane processes,
Koninklijke/Shell Laboratorium Amsterdam, 1988.

The membrane consists of polypropene with a non-swollen
thickness of 6 μm. At the beginning of an experiment there is
pure hexane (1) on the left (o) side of the membrane and
iso-octane (2) on the right (δ) side. We number the membrane
as component (3).

We consider the conditions at the beginning of the
experiment. The following parameters can be derived from the
data given:

swollen film thickness $\delta := 7.66 \cdot 10^{-6}$ m

concentrations in membrane

$$c1o := 1677 \qquad c2o := 0 \qquad \text{mol/m3}$$
$$c1\delta := 0 \qquad\quad c2\delta := 1192$$

fluxes $N1 := 7.57 \cdot 10^{-2} \quad N2 := -2.74 \cdot 10^{-2}$ mol/ (m2.s)

free solution diffusivity
$$D12_o := 6 \cdot 10^{-9} \qquad \text{m2/s}$$

membrane diffusivity (guessed)
$$D12 := 2 \cdot 10^{-9} \qquad \text{m2/s}$$

Using these values, estimate the membrane diffusivities D13
and D23. The estimate of D12 above is rather uncertain. Would
this be important?

Solution

The velocities are:
$$v1 := \frac{N1}{0.5 \cdot c1o} \qquad v1 = 9.028 \cdot 10^{-5}$$

$$v2 := \frac{N2}{0.5 \cdot c2\delta} \qquad v2 = -4.597 \cdot 10^{-5}$$

and
$$k12 := \frac{D12}{\delta} \qquad k12 = 2.611 \cdot 10^{-4}$$

The MS equations read:

$$-2 \approx 0.5 \cdot \frac{v2 - v1}{k12} + \frac{-v1}{k13} \qquad\qquad 2 \approx 0.5 \cdot \frac{v1 - v2}{k12} + \frac{-v2}{k23}$$

or

$$k13 := \left[0.5 \cdot \frac{v2 - v1}{k12} + 2 \right]^{-1} \cdot v1 \qquad k13 = 5.191 \cdot 10^{-5}$$

$$k23 := \left[0.5 \cdot \frac{v1 - v2}{k12} - 2 \right]^{-1} \cdot v2 \qquad k23 = 2.644 \cdot 10^{-5}$$

$$D13 := k13 \cdot \delta \qquad D13 = 3.977 \cdot 10^{-10}$$

$$D23 := k23 \cdot \delta \qquad D23 = 2.025 \cdot 10^{-10}$$

A check with values of D12 differing by a factor of two with the original guess shows that this parameter is not very important. It looks as if the friction between the transferring components and the membrane dominates over the liquid-liquid friction.

27. A DIALYSIS MODULE

This problem is very similar to that in Figs. 14.1-14.3. In one respect, however, it is more realistic. The compositions on both sides of the membrane are not given; you only know the compositions of the two feed streams.

In the module below, both sides are assumed to be well mixed:

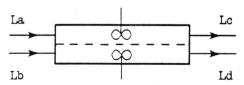

There are two transferring components: water (1) and urea (2). The transfer coefficients through the membrane are:

$$k12 := 0.3 \cdot 10^{-6} \qquad k1 := 0.1 \cdot 10^{-6} \qquad k2 := 0.02 \cdot 10^{-6} \qquad m \ s\text{-}1$$

The last two are membrane coefficients. The area and total concentration in the membrane are:

$$A := 1 \ m2 \qquad c := 3 \cdot 10^{4} \quad mol \ m\text{-}3$$

The feed flows in mol/s and the mole fractions are:

$$La := 0.0005 \qquad Lb := 0.005$$
$$x1a := 0.980 \qquad x2a := 0.020$$
$$x1b := 1.000 \qquad x2b := 0.000$$

Calculate the outgoing flows.

Hint. This problem is a little easier if you write the transport equations in terms of fluxes instead of velocities. Your unknowns are then: x1, x2, x1c, x2c, x1d, x2d, N1 and N2. So you will need eight equations.

Solution

The equations required are given in the solve block below.

initial values   x1 := 0.99   x2 := 0.01
                   x1c := 0.99   x2c := 0.01
                   x1d := 0.99   x2d := 0.01
                   N1 := 0   N2 := 0

Given

$$x1 \approx 0.5 \cdot (x1c + x1d)$$ average mole fractions
$$x2 \approx 0.5 \cdot (x2c + x2d)$$

$$(x1d - x1c) \approx \left[\frac{x1 \cdot N2 - x2 \cdot N1}{k12 \cdot c}\right] + \frac{-N1}{k1 \cdot c}$$ transport relations in terms of fluxes

$$(x2d - x2c) \approx \left[\frac{x2 \cdot N1 - x1 \cdot N2}{k12 \cdot c}\right] + \frac{-N2}{k2 \cdot c}$$

$$x2c \approx x2a - \frac{N2 \cdot A}{La} \qquad x1c \approx 1 - x2c$$ compositions from mass balances and mole fraction summations

$$x2d \approx x2b + \frac{N2 \cdot A}{Lb} \qquad x1d \approx 1 - x2d$$

$$\begin{bmatrix} x1 \\ x2 \\ x1c \\ x2c \\ x1d \\ x2d \\ N1 \\ N2 \end{bmatrix} := Find(x1,x2,x1c,x2c,x1d,x2d,N1,N2)$$

$$\begin{bmatrix} x1 \\ x2 \\ x1c \\ x2c \\ x1d \\ x2d \\ N1 \\ N2 \end{bmatrix} = \begin{vmatrix} 0.994522 \\ 0.005478 \\ 0.99005 \\ 0.00995 \\ 0.998995 \\ 0.001005 \\ -0.000025 \\ 0.000005 \end{vmatrix}$$

The flows leaving follow from:

$$L1c := La \cdot x1a - N1 \cdot A \qquad L1c = 0.000515$$
$$L2c := La \cdot x2a - N2 \cdot A \qquad L2c = 0.000005$$
$$L1d := Lb \cdot x1b + N1 \cdot A \qquad L1d = 0.004975$$
$$L2d := Lb \cdot x2b + N2 \cdot A \qquad L2d = 0.000005$$

28. PERVAPORATION OF n-C7 AND i-C8.

A set of experiments is given below on pervaporation of the
hydrocarbons n-heptane (1) and i-octane (2) through a
polyethene membrane. They are taken from J.G.A. Bitter,
Transport mechanisms in membrane separation processes,
Koninklijke/Shell Laboratorium Amsterdam, 1988.

The tables contain:
- the molar fluxes N1 and N2 and
- the molar concentrations of the two components c1 and c2
just inside the upstream membrane interface.

A few remarks on the data. The temperature is 25 C. There is
a pressure difference of one bar; the downstream pressure is
zero. The membrane has a thickness of 10.2 μm; the swelling
is very small. The concentrations in the membrane have been
calculated from a separate equilibrium model, which is based
on data taken under similar conditions.

You are to estimate the MS mass transfer coefficients and
diffusivities in this membrane.

$i := 1 ..8$

The 1-2 diffusivity has a value of roughly one third of the free solution value:

$$D12 := 1 \cdot 10^{-9}$$

The membrane has a thickness:

$$\delta := 10.2 \cdot 10^{-6}$$

$N1_i :=$	$N2_i :=$	$c1o_i :=$	$c2o_i :=$
0.000000	0.000268	0	297
0.000033	0.000263	23	290
0.000111	0.000293	64	277
0.000285	0.000250	169	238
0.000677	0.000198	339	165
0.001112	0.000108	538	83
0.001388	0.000025	666	19
0.001478	0.000000	707	0

Hint: when you calculate the driving force, you must realize
that the downstream concentrations in the membrane are zero.

Solution

The average concentrations and velocities in the membrane
are:

$$c1_i := 0.5 \cdot c1o_i \qquad\qquad c2_i := 0.5 \cdot c2o_i$$

$$ax1_i := \frac{c1_i}{c1_i + c2_i} \qquad\qquad ax2_i := \frac{c2_i}{c1_i + c2_i}$$

$$v1_i := \frac{N1_i}{c1_i} \qquad\qquad v2_i := \frac{N2_i}{c2_i}$$

The 1-2 coefficient is equal to $\qquad k12 := \dfrac{D12}{\delta}$

The driving forces have the maximum value of -2. The membrane coefficients then follow from the MS relations:

$$-2 := x2 \cdot \frac{v2 - v1}{k12} + \frac{-v1}{k13} \quad\square \qquad -2 := x1 \cdot \frac{v1 - v2}{k12} + \frac{-v2}{k23} \quad\square$$

or

$$k13_i := \left[ax2_i \cdot \frac{v2_i - v1_i}{k12} + 2 \right]^{-1} \cdot v1_i \qquad D13_i := k13_i \cdot \delta$$

$$k23_i := \left[ax1_i \cdot \frac{v1_i - v2_i}{k12} + 2 \right]^{-1} \cdot v2_i \qquad D23_i := k23_i \cdot \delta$$

$k13_i \cdot 10^6$	$k23_i \cdot 10^6$	$D13_i \cdot 10^{12}$	$D23_i \cdot 10^{12}$
0	0.902	0	9.204
1.442	0.907	14.708	9.247
1.744	1.056	17.79	10.775
1.693	1.048	17.267	10.686
2.002	1.193	20.424	12.173
2.069	1.292	21.105	13.183
2.085	1.306	21.262	13.32
2.091	0	21.323	0

To allow plotting we add a small positive number:

$$D13_i := D13_i + 10^{-14}$$

$$D23_i := D23_i + 10^{-14}$$

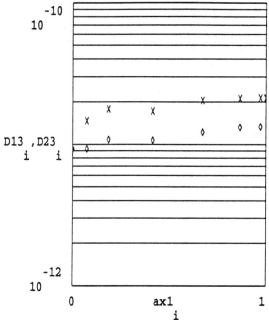

There is some variation of the diffusivities with the
composition of the feed, but the variation is not large. Do
realize that these are average values: it is quite probable
that there are variations over the membrane.

29. THE LIMITING CURRENT IN ELECTRODIALYSIS

The current through an electrodialysis membrane is limited by
polarization. The phenomena are very similar to those in
electrolysis (Fig. 11.5). There is one difference: in
electrodialysis there is also a flux of water through the
"electrode". The magnitude of this flux is determined by what
happens in the membrane.

You are to determine the magnitude of the limiting current
for the membrane in Fig. 15.5. For this purpose you should
analyse the transport in the film in the diluate compartment.
There are three components: Na+ (1), Cl- (2) and water (3).
The solution is dilute, so only the 1-3 and 2-3 interactions
are important. Also transport of Cl- through the membrane is
neglected. The diffusivities and the film thickness are:

$$D13 := 1.3 \cdot 10^{-9} \qquad D23 := 2.0 \cdot 10^{-9} \qquad \delta := 3 \cdot 10^{-5}$$

The solution has a bulk composition:

$$c := 55 \cdot 10^{3} \qquad x1o := 2 \cdot 10^{-4}$$

You may assume that the ratio of the Na+ and water fluxes is
the same as that in Fig. 15.5:

$$r13 := \frac{4.82}{23.6} \qquad r13 = 0.204$$

Go ahead.

Solution

At the limiting current the electrolyte concentration at the membrane is zero. The composition values we need are then:

x1 := 0.5·x1o   x2 := x1   x3 := 1 - x1 - x2
dx1 := -x1o   dx3 := 2·x1o

The mass transfer coefficients are:

$$k13 := \frac{D13}{\delta} \qquad k23 := \frac{D23}{\delta}$$

There are three unknowns: the velocities of the two transferring components and the potential difference. We use as starting values:

v1 := 0   v3 := 0   Dϕ := -0.1

We neglect the friction between the ionic species. The defining equations are then:

Given

$$\frac{dx1}{x1} + 40 \cdot D\phi \approx x3 \cdot \begin{bmatrix} -v1 \\ \overline{k13} \end{bmatrix} \qquad \frac{dx3}{x3} \approx x1 \cdot \frac{v1 - v3}{k13} + x2 \cdot \begin{bmatrix} -v3 \\ \overline{k23} \end{bmatrix}$$

v1·x1 ≈ 0.204·v3·x3

$$\begin{bmatrix} v1 \\ v3 \\ D\phi \end{bmatrix} := Find(v1,v3,D\phi) \qquad \begin{bmatrix} v1 \\ v3 \\ D\phi \end{bmatrix} = \begin{bmatrix} 0.00017351 \\ 0.00000009 \\ -0.05008096 \end{bmatrix}$$

Note the extremely low value of the water velocity. The current per unit area becomes:

$$F := 96 \cdot 10^{3} \qquad I := F \cdot c \cdot v1 \cdot x1 \qquad I = 92$$

The differential equations describing this system are not very difficult to solve. You will find that they give exactly the same value for the limiting current. There is one difference however. There is no fixed value of Dϕ for which the current reaches a maximum. The limiting current is approached asymptotically as the potential difference increases.

For the potential difference calculated above , the "exact" solution gives a current of 79 A/m2.

30. ELECTRODIALYSIS: HIGHER CONCENTRATIONS

Electrodialysis is a process in which electrolytes are
concentrated by the application of an electrical field. In
this tutorial we will calculate the velocities through a
membrane with a fixed negative charge. The electrolyte NaCl
is moved from the diluate to the concentrate side. The
problem is similar to that covered in Figs. 15.2-15.5. There
is one important difference: we consider higher
concentrations. Because of this we will take the penetration
of the co-ion Cl- into the membrane into account.

There are four species involved: H2O (1), Na+ (2), Cl-
(3) and the membrane or HSO4- (4).

Parameters required are:

diluate concentrations	csd := 0.055	mol NaCl 1-1
	ctd := 55	mol 1-1 total
membrane c's	cnm := 2	mol HSO4- 1-1
	ctm := 15	mol 1-1 total
concentrate c's	csc := 1.1	mol NaCl 1-1
	ctc := 55	mol 1-1 total

$$\text{membrane thickness} \quad \delta := 2 \cdot 10^{-4} \quad m$$

$$\text{diffusivities in} \quad D12 := 1.3 \cdot 10^{-10} \quad D13 := 2.0 \cdot 10^{-10}$$
$$\text{the membrane} \qquad\qquad\qquad\quad D14 := 2.0 \cdot 10^{-10} \quad D23 := 2.0 \cdot 10^{-11}$$
$$D24 := 1.3 \cdot 10^{-11} \quad D34 := 2.0 \cdot 10^{-11}$$

$$\text{potential difference} \quad D\phi := -0.2$$

Calculate the velocities and fluxes of Na+, Cl- and H2O
through the membrane.

Solution

First we calculate all mole fractions in the diluate (d) and
concentrate (c) phases:

$$x2d := \frac{csd}{ctd} \qquad x3d := \frac{csd}{ctd} \qquad x1d := 1 - x2d - x3d$$

$$x2c := \frac{csc}{ctc} \qquad x3c := \frac{csc}{ctc} \qquad x1c := 1 - x2c - x3c$$

Then we calculate the driving forces from the outside
conditions:

$$dx1 := x1c - x1d \qquad x1 := 0.5 \cdot (x1c + x1d)$$
$$dx2 := x2c - x2d \qquad x2 := 0.5 \cdot (x2c + x2d)$$
$$dx3 := x3c - x3d \qquad x3 := 0.5 \cdot (x3c + x3d)$$

$$df1 := \frac{dx1}{x1} \qquad df2 := \frac{dx2}{x2} + 40 \cdot D\phi \qquad df3 := \frac{dx3}{x3} - 40 \cdot D\phi$$

Then the mole fractions in the two interfaces of the membrane. These are the diluate (o) and concentrate (δ) sides. The co-ion fractions follow from the Donnan relation (Fig. 15.3):

$$x4 := \frac{2}{15}$$

$$x3o := \frac{x3d^2}{x4} \qquad\qquad x3\delta := \frac{x3c^2}{x4}$$

The fraction of positive ions is equal to the sum of the two negative fractions:

$$x2o := x3o + x4 \qquad x2\delta := x3\delta + x4$$

and water forms the rest:

$$x1o := 1 - x2o - x3o - x4 \qquad x1\delta := 1 - x2\delta - x3\delta - x4$$

The average values are:

$$x1 := 0.5 \cdot (x1o + x1\delta)$$
$$x2 := 0.5 \cdot (x2o + x2\delta) \qquad x2 = 0.135$$
$$x3 := 0.5 \cdot (x3o + x3\delta) \qquad x3 = 0.002$$

Now the mass transfer coefficients:

$$k12 := \frac{D12}{\delta} \qquad k13 := \frac{D13}{\delta} \qquad k14 := \frac{D14}{\delta}$$

$$k23 := \frac{D23}{\delta} \qquad k24 := \frac{D24}{\delta} \qquad k34 := \frac{D34}{\delta}$$

The velocities then follow from:

initial estimates   v1 := 0   v2 := 0   v3 := 0

Given

$$df1 \approx x2 \cdot \frac{v2 - v1}{k12} + x3 \cdot \frac{v3 - v1}{k13} + x4 \cdot \frac{-v1}{k14}$$

$$df2 \approx x1 \cdot \frac{v1 - v2}{k12} + x3 \cdot \frac{v3 - v2}{k23} + x4 \cdot \frac{-v2}{k24}$$

$$df3 \approx x1 \cdot \frac{v1 - v3}{k13} + x2 \cdot \frac{v2 - v3}{k23} + x4 \cdot \frac{-v3}{k34}$$

$$\begin{bmatrix} v1 \\ v2 \\ v3 \end{bmatrix} := Find(v1,v2,v3) \qquad \begin{bmatrix} v1 \\ v2 \\ v3 \end{bmatrix} \cdot 10^{6} = \begin{vmatrix} 1.625 \\ 2.506 \\ -1.537 \end{vmatrix} \qquad m \ s\text{-}1$$

The fluxes are:

```
N1 := v1·x1·ctm
N2 := v2·x2·ctm
N3 := v3·x3·ctm
```

$$\begin{bmatrix} N1 \\ N2 \\ N3 \end{bmatrix} \cdot 10^{6} = \begin{vmatrix} 17.806 \\ 5.068 \\ -0.035 \end{vmatrix} \qquad mol \ m\text{-}2 \ s\text{-}1$$

In reality you may expect a higher flux of the co-ion. This is because the simple Donnan relation underestimates the co-ion penetration. An important reason is microscopic inhomogeneity of the membrane.

31. POLARIZATION IN HYPERFILTRATION AND ULTRAFILTRATION

The question is: why is polarization much more important in
ultrafiltration than in hyperfiltration (reverse osmosis)?

To answer this question you need the following figures
5.3 polarization,
9.4 diffusivities in dilute liquid solutions,
16.2 reverse osmosis (hyperfiltration), fluxes,
16.9 ultrafiltration, fluxes.

Hint: make an estimate of the ratio $(v2/k12)$ in Fig. 5.3 for
the two systems. Note that $v2$ is the water velocity: this is
not always so in the subsequent figures. For this purpose you
need the thickness of the polarization film: take a value of
$\delta := 30 \ \mu m$. Also you need the diffusivities of salt and
proteins in the film; you may estimate these from Fig. 9.4.
You can obtain orders of magnitude of the water fluxes and
velocities from the other figures. You need the velocity in
the film of course, not that in the membrane!

Solution

My estimates yield:	reverse osmosis	ultrafiltration
salt and protein D's:	$D12 := 10^{-9}$	$D12' := 10^{-10}$

transfer coefficients with
$\delta := 30 \cdot 10^{-6}$

$$k12 := \frac{D12}{\delta} \qquad k12' := \frac{D12'}{\delta}$$

a UF flux mainly due to
viscous flow: the velocity
is 10 x higher in the pores.

$$u := 1.25 \cdot 10^{-4}$$
$$v2' := 0.1 \cdot u$$

HF velocity from the flux
and water concentration:

$$N2 := 0.3$$
$$c2 := 55 \cdot 10^{3}$$
$$v2 := \frac{N2}{c2}$$

so finally:

$$\frac{v2}{k12} = 0.164 \qquad \frac{v2'}{k12'} = 3.75$$

As you can see in Fig. 5.3, this means a dramatic increase in
the polarization. In reverse osmosis the effect of
polarization is minor, whereas in ultrafiltration the
membrane flux is all too often limited by the formation of
precipitates (gels) in the polarization layer.

32. A REVERSE-OSMOSIS MEMBRANE

This exercise concerns the reverse-osmosis membrane treated in Figs. 16.1-16.3. The parameters are:

upstream compositions $x1o := 0.98$ $x2o := 0.02$

downstream compositions $x1\delta$ and $x2\delta$ unknown

transport coefficients $k1 := 10^{-3}$ $k2 := 10^{-5}$

distribution coeffs $m1 := 0.1$ $m2 := 0.01$

pressure constants $p\%1 := 140 \cdot 10^{6}$ $p\%2 := 30 \cdot 10^{6}$

total liquid concentration $c := 55 \cdot 10^{3}$

temperature parameters $T := 293$ $R := 8.314$

You are to investigate the effect of the pressure difference dp across the membrane. Vary dp between -0.1 and -10 MPa. This includes the area below the osmotic pressure of the solution. Assume that the downstream composition is determined by the values of the salt and water fluxes through the membrane.

Investigate the effect of dp on the retention of the membrane. The retention can be calculated from:

$$r := 1 - \frac{x2\delta}{x2o} \quad \square$$

Solution

The downstream compositions are unknown. They are determined by the fluxes through the membrane, which are also unknown. The defining equations are shown in a solve block below.

Pressure difference $dp := -10 \cdot 10^{6}$

Initial values $x1d := 1.00$ $x2d := 0.0001$
                  $v1 := k1 \cdot 0.05$ $v2 := 2 \cdot k2$
                  $N1 := v1 \cdot 5000$ $N2 := v2 \cdot 10$

Given

$$2 \cdot \frac{x1d - x1o}{x1d + x1o} + \frac{dp}{p\%1} \approx \frac{-v1}{k1} \qquad 2 \cdot \frac{x2d - x2o}{x2d + x2o} + \frac{dp}{p\%2} \approx \frac{-v2}{k2}$$

$$N1 \approx v1 \cdot m1 \cdot c \cdot 0.5 \cdot (x1d + x1o) \qquad N2 \approx v2 \cdot m2 \cdot c \cdot 0.5 \cdot (x2d + x2o)$$

$$x1d + x2d \approx 1$$

$$\frac{x1d}{x2d} \approx \frac{N1}{N2}$$

$$\text{Find}(v1,v2,N1,N2,x1d,x2d) = \begin{vmatrix} 0.000052 \\ 0.000022 \\ 0.281308 \\ 0.000126 \\ 0.999551 \\ 0.000449 \end{vmatrix} \begin{matrix} v1 \\ v2 \\ N1 \\ N2 \\ x1d \\ x2d \end{matrix}$$

With this program I have carried out a number of runs. These
did require several adjustments to the starting values to
obtain convergence. The results are plotted below, using
logarithmic scales.

As you can see there are two very different regimes. Small
pressure differences do not provide much separation; the
fluxes of the water and salt are proportional to their mole
fractions. Only at pressure differences in excess of the
osmotic pressure of the feed ($\pi 1$) is a proper separation
observed. The water flux increases very rapidly around this
point.

$$i := 0 \, ..16 \quad . \quad x2o := 0.02$$

dp_i :=	$N1_i$:=	$N2_i$:=	$x2\delta_i$:=
-10	281.3	.1260	0.0004
-8	204.3	.1220	0.0006
-6	128.2	.1170	0.0009
-4	55.96	.1075	0.0019
-3	26.45	.0965	0.0036
-2	9.493	.0731	0.0076
-1.5	5.432	.0568	0.0103
-1.2	3.727	.0462	0.0122
-1.0	2.841	.0389	0.0135
-0.8	2.091	.0313	0.0147
-0.6	1.451	.0236	0.0160
-0.4	0.900	.0159	0.0174
-0.3	0.652	.0120	0.0181
-0.2	0.420	.0080	0.0187
-0.15	0.309	.0060	0.0190
-0.12	0.246	.0048	0.0191
-0.10	0.203	.0040	0.0193

dp_i ▫ in MPa

$N1_i , N2_i$ ▫
in mmol m-2 s-1

$$r_i := 1 - \frac{x2\delta_i}{x2o}$$

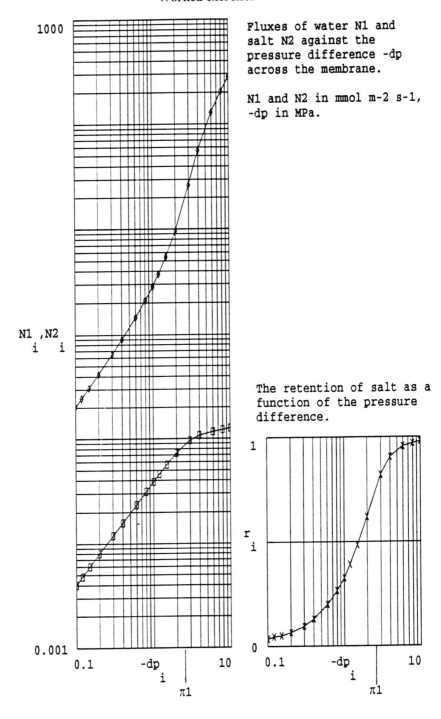

Fluxes of water N1 and salt N2 against the pressure difference -dp across the membrane.

N1 and N2 in mmol m-2 s-1, -dp in MPa.

The retention of salt as a function of the pressure difference.

33. ULTRAFILTRATION

Have a look at Fig. 16.9. There are several transport mechanisms for water and for the protein:
- diffusive transport (due to activity and pressure gradients) and
- viscous flow.
Which is the most important?

Take only the most important mechanism into account. Calculate the retention of the membrane for spherical proteins with different diameter ratios.

Solution

The main transport mechanism for both species is viscous flow. The fluxes are:

$$N1 := u \cdot x1 \cdot c \quad\quad\quad N2 := u \cdot x2 \cdot c$$

and the downstream fraction protein is:

$$x2'' := \frac{N2}{N1 + N2} \quad \text{or} \quad x2'' := \frac{x2}{x1 + x2}$$

This is practically equal to x2. The retention is:

$$R := 1 - \frac{x2''}{x2'} \quad \text{or} \quad R := 1 - \frac{x2}{x2'}$$

The last ratio is determined by geometrical exclusion (see Fig. 16.5). The retention then becomes:

$$i := 0 \,..20$$

$$r_i := \frac{i}{20} \quad\quad\quad R_i := 1 - \left[1 - r_i\right]^2$$

$$\text{with} \quad\quad\quad r_i := \left[\frac{d2}{d3}\right]_i$$

A better model, which takes the parabolic velocity profile in the pore into account, yields:

$$R'_i := 2 \cdot r_i^2 - r_i^4$$

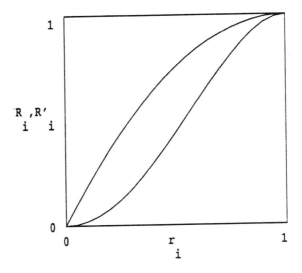

R above;
R' below.

34. TRANSPORT PARAMETERS OF A MEMBRANE

You will not find membrane parameters in the form needed for
MS calculations in the data sheets of membrane manufacturers.
The reasons for this are:
- Membrane manufacturers are not yet aware of the advantages
  of the MS description and
- MS parameters are difficult to determine.
Even so, engineers with a Sherlock Holmes attitude can get a
lot of information out of product and application
descriptions.

As an example, we have a look at the Cuprophan dialysis
membranes produced by AKZO. These are hollow-fibre cellulose
membranes.

The following are available:
- a data sheet,
- a sheet containing Fick diffusivities of some components
  through the membrane,
- a sheet with sieving coefficients,
- a short description of a dialysis process for reducing the
  alcohol content of beer. This uses Diatherm membranes, but
  these are said to be very similar to Cuprophan.
The important parts are summarized below.

From these data you would like to obtain an idea of:
* the geometry of the wet fibres:
  - inner diameter do,
  - wall thickness δ.
* the structure of the wet fibre wall:
  - void (water) fraction ε,
  - pore diameter d3.
* the transport coefficients:
  - water / ethanol,
  - water / membrane and
  - ethanol / membrane.

We suggest that you try to do this using the parallel-pore
model of ultrafiltration in Chapter 16. You can make this
model slightly more realistic by allowing pores with a
tortuosity. Ethanol takes the place of the protein. The
ethanol concentrations are low.

Properties of CUPROPHAN hollow-fibre membranes.

inner diameter	dry	200 μm
wall thickness	dry	11 μm
outer diameter	wet	270 μm
ultrafiltration flux		4 ml/(h*m2*mm Hg)
water content	dry?	10 %
glycerol content	dry	9.3 %

diffusivities

species	molar mass (g mol-1)	Df (10^-10 m2 s-1)
ethanol	46	1.13
lactose	342	0.24
urea	60	1.35
glucose	180	0.35
oxygen	32	2.04

sieving coefficients

species	diameter (nm)	S
urea	0.8	1.00
vitamin B12	1.6	0.88
inuline	2.8	0.25
HS-albumin	6	0.00

process parameters for alcohol reduction in beer:

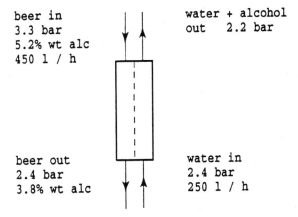

beer in
3.3 bar
5.2% wt alc
450 l / h

water + alcohol
out   2.2 bar

beer out
2.4 bar
3.8% wt alc

water in
2.4 bar
250 l / h

Solution

Be warned that this is not much more than a set of (we hope) intelligent guesses.

Geometry

From the data sheet we find:

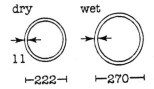

dry   wet

11

├222┤   ├270┤

We assume that the geometry remains the same on swelling. For the wet thickness we then obtain:

$$\delta := \frac{270}{222} \cdot 11 \cdot 10^{-6}$$

$$\delta = 1.338 \cdot 10^{-5}$$

Water content

The dry value is 10%. In addition the membrane contains 9%
glycerol which is presumably displaced by water. So the
volume fraction of aqueous components in the dry membrane is
about 19%. In the sketches below, we have rolled out the
cross- sections of the dry and wet membranes and indicated
the volumes of the polymer and water. Again we have assumed
no change in the relative dimensions on swelling and no
change of the actual polymer volume.

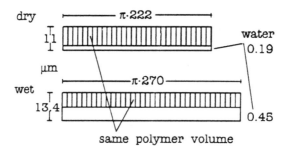

For the wet fibre we find $\varepsilon := 0.45$ (We are told that
this estimate is somewhat low.)

Pore diameter

The sieving coefficients are related to the retention:

$S := 1 - R$ ▫

As you can see the retention is almost complete for molecules
with a diameter of around 4 nm. This suggests:

$d3 := 4 \cdot 10^{-9}$

You can check this result with the ultrafiltration flux in
the data sheet. This is actually the ratio of the volumetric
flux of pure water and the corresponding pressure difference:

$$\frac{u \cdot \varepsilon}{-dp} := 4.0 \cdot \frac{ml}{m^2 \cdot hr \cdot mmHg} \quad ▫ \quad or \quad \frac{u \cdot \varepsilon}{-dp} := 8.4 \cdot 10^{-12} \quad ▫ \ (SI\ units)$$

The corresponding diameter can be deuced from the Poiseuille
equation:

$$d3^2 := 32 \cdot \frac{u \cdot \delta \cdot \mu}{-dp} \quad ▫ \quad yielding \quad d3 := \sqrt{\frac{32 \cdot 8.4 \cdot 10^{-12} \cdot \delta \cdot \mu}{\varepsilon}}$$

$$d3 = 2.827 \cdot 10^{-9}$$

This is in the right range. A pore diameter of 4 nm is obtained from this relation if a tortuosity is assumed:

$$\tau := \sqrt{\frac{4.0}{2.837}} \qquad \tau = 1.187 \qquad \text{(which is not unreasonable)}$$

Transport coefficients

We estimate the diffusivities using Figs. 16.5, 16.6 and 16.7. The molecular diameters are taken as:

(1) water $d1 := 0.4 \cdot 10^{-9}$

(2) ethanol $d2 := 0.5 \cdot 10^{-9}$

(3) pores $d3 := 4 \cdot 10^{-9}$

The Fick diffusivity of ethanol in water was found in a handbook:

$$Df2 := 0.85 \cdot 10^{-9}$$

The MS coefficient in our model becomes

$$D12 := \frac{Df2}{\tau^2} \qquad D12 = 6.029 \cdot 10^{-10}$$

The water - membrane coefficient is:

$$D1 := \frac{R \cdot T}{8 \cdot V1} \cdot \frac{d1 \cdot d3}{\mu13 \cdot \tau^2} \qquad D1 = 1.728 \cdot 10^{-9}$$

Because we do not really know $\mu13$, the uncertainty in this value is large. A rough check can be obtained as follows. In the description of the dialysis of beer it is claimed that no dilution or concentration of beer occurs. In the process description it is seen that a small pressure difference is applied between the two sides of the membrane. The pressure difference varies across the module, but on average has a value of about 0.5 bar. This value is probably required to counterbalance the diffusive flux of the water into the beer.

The main terms in the diffusion equation of water are:

$$\frac{dx1}{x1} := -v1 \cdot \frac{\delta \cdot \tau^2}{D1} \quad \square \text{ or } \quad v1 := \frac{-D1}{\delta \cdot \tau^2} \cdot dx1 \quad \square$$

The viscous velocity is

$$u := \frac{-1}{32} \cdot \frac{dp \cdot d3^2}{\mu \cdot \delta \cdot \tau^2} \quad \Box$$

With   dx1 := 0.01   and   dp := $-5 \cdot 10^4$ the water fluxes cancel when

$$D1 := \frac{-1}{32} \cdot \frac{dp \cdot d3^2}{\mu \cdot dx1} \qquad D1 = 2.5 \cdot 10^{-9}$$

which is surprisingly close to our first estimate.

For the alcohol - membrane coefficient we use Fig. 16.7:

$$\frac{d2}{d3} = 0.125 \quad \text{so} \quad \frac{k12}{k2} := 0.7 \; \Box \quad \text{and} \quad D2 := \frac{D12}{0.7}$$

so

$$D2 = 8.612 \cdot 10^{-10}$$

The only bit of experimental information available on diffusivities is the Fick diffusivity of ethanol through the membrane:

$$Df2 := 1.13 \cdot 10^{-10}$$

The transport relation for ethanol under such diffusion measurements is:

$$\frac{dx2}{x2} := -v2 \cdot \frac{\delta}{D12} - v2 \cdot \frac{\delta}{D2} \quad \Box$$

So

$$Df2 := \frac{\varepsilon}{\frac{2}{\tau}} \cdot \left[\frac{1}{D12} + \frac{1}{D2} \right]^{-1}$$

and our estimates of D12 and D2 yield:

$$Df2 = 1.132 \cdot 10^{-10}$$

(By coincidence, this is almost identical to the experimental value.)

All these parameters look fairly consistent. This is of course no guarantee that the membrane really looks like our reconstruction.

Parameters used throughout the calculations are:

$$R \equiv 8.314 \qquad T \equiv 293 \qquad \mu \equiv 10^{-3}$$

$$V1 \equiv 2 \cdot 10^{-5} \qquad \mu13 \equiv 10^{-2}$$

35. ION EXCHANGE: A BATCH EXPERIMENT

In a batch experiment, 1 l of a dilute HCl solution comes into contact with a large excess of cation exchanger beads. The ion exchanger beads are initially in the Na+ form and have a total surface area of 5 m2. It may be assumed that there is no net water or chloride flux between the ion exchanger and the solution.

There are four species in the liquid: Na+ (1), H+ (2), Cl- (3) and H2O (4). However only the 1-4 and 2-4 frictional interactions are important in the liquid. The mass transfer coefficients are:

$$k14 := 1.3 \cdot 10^{-5} \qquad k24 := 9.3 \cdot 10^{-5}$$

The total electrolyte concentration and the mole fractions in the bulk of the liquid at the beginning of the experiment are:

$$c := 55 \cdot 10^{3} \qquad x1\delta := 0 \cdot 10^{-4}$$
$$x2\delta := 2 \cdot 10^{-4}$$
$$x3\delta := 2 \cdot 10^{-4}$$

Under the above conditions, the main mass transfer resistance lies in the liquid film. Calculate the initial flux of sodium.

Also, do the same calculation for ion exchange particles initially in the H+ form, which are brought into contact with a dilute solution of NaCl.

Solution

The following equations describe the system:

initial guesses
$$x1o := 0.0003 \qquad v1 := -10^{-5}$$
$$x2o := 0.0000 \qquad v2 := 10^{-5} \qquad d\phi := 0$$
$$x3o := 0.0003$$

given

$$dx1 := x1\delta - x1o \qquad x1 := 0.5 \cdot (x1\delta + x1o)$$
$$dx2 := x2\delta - x2o \qquad x2 := 0.5 \cdot (x2\delta + x2o)$$
$$dx3 := x3\delta - x3o \qquad x3 := 0.5 \cdot (x3\delta + x3o)$$

$$x1o \approx x3o \qquad \text{electroneutrality}$$

$$\frac{dx1}{x1} + 40 \cdot d\phi \approx \frac{-v1}{k14} \qquad \text{transport relations}$$

$$\frac{dx2}{x2} + 40 \cdot d\phi \approx \frac{-v2}{k24}$$

$$\frac{dx3}{x3} - 40 \cdot d\phi \approx 0$$

$$v1 \cdot x1 + v2 \cdot x2 \approx 0 \qquad \text{no current}$$

$$\begin{bmatrix} x1o \\ x3o \\ v1 \\ v2 \\ d\phi \end{bmatrix} := \text{find}(x1o, x3o, v1, v2, d\phi) \qquad \begin{bmatrix} x1o \\ x3o \\ v1 \\ v2 \\ d\phi \end{bmatrix} = \begin{vmatrix} 0.000535 \\ 0.000535 \\ 0.000038 \\ -0.000101 \\ -0.022787 \end{vmatrix}$$

$$N1 := v1 \cdot (0.5 \cdot (x1o + x1\delta)) \cdot c \qquad N1 = 0.000557$$

The same calculation, but with Na+ and H+ on opposite sides of the film, yields:

$$N1 = -0.000211$$

The fluxes are almost three times lower. This has indeed been found by experiment.

36. SURFACE DIFFUSION IN A MICROPOROUS PLUG

Pope, (see Trans.Farad.Soc. 63, 734 (1967)), has published
data on surface diffusivity (Fick diffusivity!) of SO_2 in a
microporous plug of Spheron 6(2700). His data, at 0 C, are
presented below in a somewhat different form using θ, the
fraction of the surface occupied by SO_2.

surface occupancy diffusivity	Fick diffusivity	tracer
	x 10-8 m2 s-1	x10-8 m2 s-1
$k := 0 \;..5$		
$\theta :=$	$Df :=$	$Dt :=$
$\overline{k}$	$\overline{k}$	$\overline{k}$
0.001	8.0	8.0
0.111	7.6	5.0
0.333	7.1	2.8
0.556	7.3	1.83
0.778	11.9	1.23
0.999	19.9	0.94

Also given in the above table are the associated measurements
of the diffusivity of labelled SO_2 in a mixture of labelled
and unlabelled species. The labelled species was present in
trace amounts. Interestingly the tracer diffusivity exhibits
a sharp decrease whereas the Fick diffusivity increases with
surface occupancy.

You are required to rationalize the experimental observations
of Pope by formulating surface diffusion in terms of the MS
diffusion equations.

Hints

There are three species: SO_2 unlabelled (1), SO_2 labelled (2)
and the solid surface (3). The data given above are Fick
diffusivities defined by:

$$N_i := -c \cdot D_i \cdot \frac{d\theta_i}{dz} \;\square$$

N_i is the surface flux and θ_i the
surface fraction of the component
considered. c is the total
concentration of sites on the surface.

You have data from two different sets of experiments. In the
first, only (1) and (3) are involved. There is a finite
surface concentration gradient of (1) and you must expect
non-ideality to play a role. In the second experiment, only a
trace of (2) is moving: the gradient in the non-ideality
should be negligible.

You may assume that adsorption of both SO_2 species is
governed by the same Langmuir equation. Also $\theta_1 \gg \theta_2$. Then:

$$y_i := K \cdot \frac{\theta_i}{1 - \theta_1} \;\square$$

where y_i is a measure of the gas
composition and K a constant.

The activity coefficient of (1) is then:

$$\tau 1 := \frac{1}{1 - \theta 1} \quad \square$$

and the gradient of the activity is then easily worked out to be:

$$\frac{d(\ln(\tau 1 \cdot \theta 1))}{dz} := \left[\frac{1}{1 - \theta 1}\right] \cdot \frac{1}{\theta 1} \cdot \frac{d\theta 1}{dz} \quad \square$$

Write out the MS transport relations for the two cases.

Solution

The transport relations for the mobile species in the two experiments are:

$$\frac{d(\ln(\tau 1 \cdot \theta 1))}{dz} := \frac{-v1}{D1} \quad \square \quad \text{and}$$

$$\frac{1}{\theta 2} \cdot \frac{d\theta 2}{dz} := \theta 1 \cdot \frac{-v2}{D12} + \frac{-v2}{D2} \quad \square$$

The fluxes follow from:

$$Ni := vi \cdot \theta i \cdot c \quad \square$$

After some rearrangement we obtain:

$$N1 := -c \cdot \left[\frac{.1}{1 - \theta 1}\right] \cdot D1 \cdot \frac{d\theta 1}{dz} \quad \square \quad \text{so} \quad D1_k := \left[1 - \theta_k\right] \cdot Df_k$$

and

$$N2 := -c \cdot \left[\frac{\theta 1}{D12} + \frac{1}{D2}\right]^{-1} \cdot \frac{d\theta 2}{dz} \quad \square \quad \text{so} \quad Dt_k := \left[\frac{\theta_k}{D12_k} + \frac{1}{D2_k}\right]^{-1} \quad \square$$

As the two species are very similar, we can take D1 = D2. This yields:

$$D12_k := \frac{\theta_k \cdot D1_k \cdot Dt_k}{D1_k - Dt_k}$$

The values of D1 (The MS coefficient describing the gas -

solid friction) and Dt (the tracer coefficient) are plotted
below.

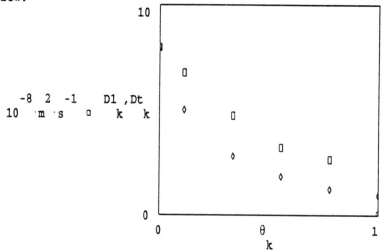

Both diffusivities decrease monotonically with increasing
surface coverage. The point in the bottom corner at the right
has little meaning; at this point the transformation from
Fick to MS coefficients contains serious inaccuracies.

It is interesting to note that the Dl's follow the
logarithmic interpolation rule for MS diffusivities quite
reasonably:

j := 0 ..1

x_j :=	y_j :=
0	8
1	1.8

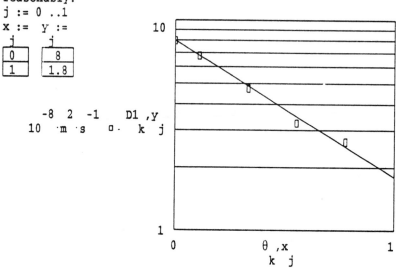

For the four intermediate points we can also calculate the
values of D12 (again in 10-8 m2 s-1). They appear to be
fairly constant.

k := 1 ..4

$D12_k$
2.135
2.281
2.337
1.791

Index

Bold numbers mean that the reference involves a figure.

Index